T0326483

The New Television Ecosystem

Participation in Broadband Society

Edited by Leopoldina Fortunati / Julian Gebhardt / Jane Vincent

Volume 7

PETER LANG
Frankfurt am Main · Berlin · Bern · Bruxelles · New York · Oxford · Wien

Alberto Abruzzese / Nello Barile / Julian Gebhardt /
Jane Vincent / Leopoldina Fortunati (eds.)

The New Television Ecosystem

PETER LANG
Internationaler Verlag der Wissenschaften

Bibliographic Information published by the Deutsche Nationalbibliothek
The Deutsche Nationalbibliothek lists this publication in the Deutsche Nationalbibliografie; detailed bibliographic data is available in the internet at http://dnb.d-nb.de.

ISSN 1867-044X
ISBN 978-3-631-61657-4

Internationaler Verlag der Wissenschaften
Frankfurt am Main 2012

www.peterlang.de

Acknowledgements

The editors would like to thank Ulrika Müller (Humboldt-University of Berlin) for her helpful contribution in the editing and proofreading of this volume.

The editors wish to thank the staff at Peter Lang GmbH, Berlin, for their support in establishing a new series on new ICT and society called "Participation in Broadband Society".

Contents

Alberto Abruzzese

Introduction

> Environments work us over and remake us. It is man who is the content of and the message of the media, which are extensions of himself. Electronic man must know the effects of the world he has made above all things (McLuhan, 1972, p.90).

I use this quotation from Marshall McLuhan as the epigraph to this volume not only as a tribute to the end of his Centenary, but more so because it expresses the notion of environment in a way that is most pertinent to media studies and to the television – the topic of this volume. In this introduction I explore the new issues and challenges facing society in which the new television ecosystem is developing by putting forward new arguments that challenge and contextualise the debate, and the discourse in this volume. McLuhan posited that:

> New technological environments are commonly cast in the moulds of the preceding technology out of the sheer unawareness of their designers (McLuhan, 1972, p. 47).

This is exemplified in the present ecosystem which has been framed and, indeed, shaped by McLuhan's legacy and is one in which the multiple facets of our global digital age are put in sharp focus via the medium of the television (TV).

McLuhan's famous motto "the medium is the message" (1964) heralded much debate, not least in his own works such as quoted here, in which he explored further the notion of environment as a system of interaction between material and nonmaterial media (technologies, words, images, sounds, people etc.). This present book follows a similar approach, as it aims to represent a realistic picture of contemporary international TV studies. Looking more broadly at television in the context of media studies one can see that the eco-systemic vision is not just an opportunistic idea that follows the same trajectory of other industrial sectors; rather it is a different paradigm that pushes us to think and use media in a different way. When the routines of old capitalism, and old technologies are critiqued from the point of view of a new ecologic thought, it is clear that we cannot just ignore this process and continue to study television as an individualistic technology. I evoke here the intellectual example of one of the most important contemporary sociologists, Edgar Morin whose debut in the sixties, L'Esprit du temps (1962), provided the academic world with a fundamental contribution to the development of the cultural industry. Even in this early work we find some examples of a systemic thought, although it was not until the 1980s, when sociologists started giving attention to the dynamics of complex systems that Morin moved in his later works to a complete conjunction with the ecologist thought. The existential and cultural trajectory we see in his work is useful in understanding the evolution of a medium (television) that is also a huge metaphor of a part of society.

Television broadcasting is, indeed, not only the main technological mode of communication in the development of mass society; it also offers an approach that informs many sectors of social life, from logistics to marketing and from advertising to fashion. We can say that TV broadcasting represents the amplification of the mass market logic from the nineteen fifties to the nineties. Now, with the twilight of this broadcasting era we can see how television is trying to re-negotiate its function and fit with other new media, a point explored by new theorists such as Bolter and Grusin (1999). This is, of course, much more visible in the general process of the digitalisation of the TV broadcasting that will eventually completely transform the way we consider TV in the near future.

McLuhan's thesis, that the old media have become a part of the new media is perhaps too enigmatic if we do not, at the same time, keep in mind another even more crucial thesis of his. According to this the development of the forms of communication – from their primordial origin up to the present regimes of meaning of their highest technological permutation – has been characterised by an ongoing conflict between the visual languages and the languages of the senses. The first of these language approaches is governed by written activity that dramatizes the images (books, the press, cinema and television); while the second is ruled by senses, by the body or more exactly by the flesh, which is violently inhibited and subjugated by social bonds (governments, strong roles and identities, armies: products of both civilization and modernization). This is a line of thought that is against both the institutions and the western idea of progress which in the 1930s – one of the most heretical times for European culture – was viewed as the conflict between knowing and not-knowing. In short, if we accept – again following McLuhan – the idea according to which technology is a prosthesis of the human body (1964), the process in which today the digital languages are emerging against the analogue ones, the society of networks against the traditional ones, personal relations against social bonds, can be interpreted as a resounding re-appropriation of ourselves in terms of the most intimate part of human nature (I must highlight that the aforesaid is absolutely not in line with the "principle of hope" of humanist thought in its current ideological versions, either extremist or moderate: human nature is *per se* an expression of both violence and suffering). It is actually true that both the major TV stories and collective information as well as post-modern TV are increasingly and rapidly migrating, moving towards the networks. The former are simultaneously traditional, generalist and deeply rooted in the culture of a country, and are domestic, for families and for developing sovereign identity; the latter – the hyper-modern TV – broadcasts segmented stories and situations, very niche ones, forming an audiovisual product distributed for the sake of private consumption and not for linear, historical and social use by the public sector and opinion forming. However, it is also true that while the two streams come together, the traditional TV contents are to be immersed in such a powerful medium that it redefines their significance, their points of view and perspectives. On the contrary, the expressive platforms of the digital networks can potentially multiply the points of view and the perspectives of communication endlessly, as never before, within the media framework of modern society. Therefore, their existence seems to contradict (potentially of course) McLuhan's assumption itself (at least the assumption from which we have started). The content of the message (the latter's statute on which modern media theories are essentially based) does not survive but on the contrary seems to dissolve into the extemporaneousness of the relational practices (some sitcoms/soaps have already given clear signs of this occurrence).

What is becoming of the world we are living in? What is the destiny, now and moreover in the future, for the social subjects in this time of transformation, suspended between a modern society and a network society? How are the political, institutional, economic and social subjects of information in those human environments becoming characterized by a higher and higher media density and therefore by connective skills eager to spread everywhere? How is the activity of the socializing agencies (which have to negotiate the meaning of the digital expressive platforms) restructuring itself within the medium and long terms of their development? How are the lives of those who – as adventurers, pioneers and finally natives – have started to inhabit the new media sphere evolving and transforming in ways that are more or less in conflict with the geopolitical models of the modern society and with the icons of the mass civilization that evolves and transforms itself? What is the meaning and the destiny of frontier wars? They have been evident since the very beginning, exploding between the creatures of the entertainment society and the creatures of a consumption society. Is it between the citizen-audience and the citizen-actor? Between the consumer and the "prosumer"? Between the external worlds of the public environment and the inner worlds of private life? These are the questions that have now arisen. These are the problems that we are currently tackling. Now, like many other times in the past, innovation has revealed itself during a catastrophic passage of eras. We are at a crossroads between the past and the future, a really momentous moment: in the middle of a systemic collapse that demolishes the certainties (and also the interests) of both the capitalist economies and the progressive ideologies and democratic politics.

This is about an epoch-making crisis, both structural and cultural. Has it exploded just before or after the move to Digital TV, to the information tools' digital languages? I personally believe that the upheaval we are facing does not concur with the modern concept of crisis which is totally dialectic, essentially affirmative and positive, i.e. necessary for our improvement, transforming our own discontinuity into continuity. On the contrary, I believe that the extreme and blatant catastrophe that our daily life is going through does not herald the promise of a revolution in the modern sense – that of short term (according to the aristocratic, bourgeois or proletarian traditions), but is actually the first signal of a mutation into the long term (similar to the millennial transition of the nomadic regimes to rural civilization). If McLuhan was able to see the computer era in television, we are supposed to see in the networks not what is evident since it is immediate, but what is not visible, thinkable nor yet happened.

In any case, what it is about is a crisis that is closely pursuant to the dizzy leap that the Internet has caused: the junctions, the plexuses and the tides of forms of relational lives that have emerged undermining – in their maximum acceleration and deceleration – even the grounds of what is Modern. It is almost as if the Old World, which is *now* past history – Civilization, Western, Westernization – has lost the languages (the contents and the forms) with which it was able to express its own power or its own distress. Therefore, the current metamorphoses of communication media – given the clash between mass media and personal media – are revealing (although still in an unclear manner) a new Tower of Babel, which, while crumbling in ruins, possesses many diverse and different languages. It is a Babel that feels nostalgic and, at the same time, recognizes the overbearing desire to rebuild the unique image of its own power.

In studying the current media situation we are asked to analyse both the past and the future of languages that are facing each other on the fields of digital networks more and more often. Is it about a war for new land between old and new social entities? Talking about new social entities could mean putting an end (or at least hoping to do so) to the tradition of the "modern ones" who assume presumptuously to be "new". It seems therefore more suitable to say that we are still within the boundaries of the traditional modern conflict between avantgarde and rearguard. We are on the battlefield, therefore, of a kind of uncertain and confused living. This is a living that is weak and precarious and, therefore, demands to make a decision – a stand, in order to divide and reassemble what exists. The materials analysed in this book, although sector specific, will hopefully contribute to orient ourselves in the current media scenario and lead to further discussion.

What are entailed by "discussion" here are the reasons provided by social entities, their apparatuses and mechanisms which justify the way in which they act. In order to debate there must be a situation, a field where choices – which are divergent and different in terms of strength and consensus – are conflicting. They are choices that divide and must be overcome by the disagreeing parties through some tactical moves that can safeguard their own strategic vision and their own goals. In those environments that are deeply immersed into the needs and interests of the government systems – markets, businesses, political parties, institutions, laws, organisations, corporations, movements – the democratic debate has to proceed through an extremely complex mass of fields of strength, which are different and hostile against each other. It is difficult to find a consistency in those processes because of the clash of interests – both material and nonmaterial, real and imaginary. The clash divides and/or unifies the economic and political environments, the money for globalisation and the bodies and places of living. Both the actors and the factors at stake are so complex that they are like an opaque screen beyond which it is very difficult to recognize the forces that really influence the transformation of a regime of meaning – and power – into an additional one.

The need to make a decision, as publicly debated as it may be, is actually the result of a series of incoherent survival instincts mixed with a dose of survival spirit. Such instincts come alive mainly motivated by self interest; as a consequence each transformation of human existence is always achieved by social action so lacking in transparency that it comes to be perceived as phenomena. The multiplicity of different subjectivities – to be aware of themselves and due to this awareness thus in conflict with each other – turns into the objectivity of a uniform, unique and subjectively inscrutable force, which, nevertheless, we call (or should call) society. It is in fact society itself that acts objectively as a filter, hiding the existence: the strict tie, which forces both roles and social professions to the division and abstraction of work (the organizational models of modern society are based on them), becomes obligation and even sense of responsibility. The public field has therefore its own hidden space, a backstage, which renders "necessarily" a kind of simulation and pretence of itself: such an existence – like appearing from the outside – that practices and social apparatuses, being themselves bound to a specific interest of their own, cannot be revealed. The subject, both as a person and as a collectivity, feels it must undergo an influence of mysterious and fatal forces: debate is no longer a free negotiation but becomes a forced negotiation: from practically open, it is confined and forced to be aware of a destiny already marked, an obligation already set to move forward. The public debate is made of scraps, marginal discussions; not about what is happening, or what is about to, but rather about what has already passed. Choices to be taken for the future have been already decided by the recent or the remote past. Therefore the present, especially in its dimension of complete media penetration, is assigned the task to carry out – inter-media fiction, symbolic ritualisation, and emotional "massage" – to represent the plots of power as destiny; they have been clouded in the coils and complexities of relations in society and have won, even before being democratically revealed, discussed and professionally processed.

In my introduction I have tried to give a general outline of the medialogical problems in order to represent a scenario to be used when asking ourselves about the techno-cultural quality of the current transition from the analogue media to the digital one, from TV languages to network ones. My considerations herein have arisen after having analysed, with quite a heretic eye, the democratic regimes in the complex systems of a post-modern society. The material gathered in this book is the product of a rightfully and advisably more cautious vision. If I have dared to introduce my post-democratic and anti-modern vision first, it is because, in any case, my vision, extremist as it is, can contribute to evaluate the critical relevance of human entry into the regimes of meaning, which we define as societies of networks (Castells, 1996). I hope my vision can allow us to move forward without coming back to the ideological dichotomies typical of the short term in revolutions. The online practices and media, along with the human element incorporated in their essence forces us to ponder about power, however, not limiting our reasoning to recall the tragically glorious heritage of the social theories in modern tradition solely related to political facts. There are actually some moments of self-awareness, such as those most deprecated by critical theories and anti-capital political thought: the place of sensory, emotional, initiatory – inner, secret, sacred – exaltation, where the hedonism of consumption and the violence of power, pleasure and sufferance of the flesh, happiness and death of the individual, come together. They meet on the same unfair and terrible field of laws of nature and society.

With the last decade of the twentieth century television has turned its aesthetics to be much closer to ways that can experiment with the contemporary social media. The Reality show, for example, is the sign of a twilight of the golden TV era that underlines two main processes: first, the way in which the audience is pulled into the heart of the representation and second, the way in which the matter of the *mise en scène* becomes the authentic dimension of everyday life. Digitalisation is not only modifying TV technologically but I would say anthropologically. This process, driving the user into the core of the production, has changed completely the meaning of the word "medium". We can say that the discovery of an experiential and emotional world around the audience has forced TV managers to make the identity and the interface of this old medium softer and more emotional. I see the confrontation between the technological determinism and the holistic vision as just an excuse to define one's own field of explanation. Both are coming from the positive sciences and both are trying to deal with the difficulties of explaining without a dramatic reduction of meaning. This is why, in our book, we try to confront the different perspectives around the general and the legitimised consideration of the centrality of the user.

Overview of this Volume

This book investigates the new cultural and social shaping of Digital Television in which the old, analogue television is being diluted. It represents the beginning of an overdue analysis and in particular it covers four main fields of research: The role of emotion in the new television ecosystem, Practices of use of Digital Television Audiences, The new types of Digital Television: Mobile Television, Neighbourhood TV and Web TV and finally Behaviour and Attitude towards Digital Television.

The first chapter of this book aims to investigate the diversification of the television audience that has taken place after the advent of the various platforms now available such as Digital Terrestrial Television, Satellite TV, Cable TV, IPTV, as well as exploire the emotional fabric that people attach to television and its new forms. Leopoldina Fortunati's and Sakari Taipale's contribution on the one hand outlines the main features of the diffusion and adoption of the new forms of television in the five most populous and industrialized countries in Europe (Italy, France, Germany, UK, and Spain) and on the other applies Russell's circumplex model to the feelings people associate with television by using the same battery of emotions that was in a previous survey in 1996.

Along similar lines, in Chapter 2, Nello Barile goes deep into the relational dynamic between technology and emotion, trying to stress how TV is turning its original mission and can now be used as a personal medium by users and also producers. Here the question is how the cultural device of "confession" (Foucault, 1978) becomes the point of contact between old mainstream TV rules and new Web 2.0 self-expression. In this development, the personal Web TV is not only a content provider but more a tool for public relations that extends the celebrity brand name to multi-existential dimension. This sophisticated storytelling about the celebrity's everyday life is the core of a strategic integration between mass and customised media that feeds a "soft" form of cultural hegemony.

Chapter 3 moves us from the lights of the stage to the darkness of a tragedy, Emiliano Treré and Manuela Farinosi decided to handle the difficult matter of the analysis of a TV cross-media platform – FromZero TV – that was created to offer a countervision exploring the difficulties of Italy's L'Aquila citizens, after the earthquake that struck the city so forcefully in 2009. The *Fromzero platform* was created to compensate the lack of information proffered by the Italian broadcasters in the coverage of the tragedy. Using qualitative methodologies (video analysis and interviews), the authors found that *FromZero TV* offered a considerate and balanced representation of sense of pain and grief of the victims of the catastrophe. This cross-media platform, in the conclusion of Treré and Farinosi represents an interesting "experiment" in how new forms of television on the Internet can offer alternative representations of events and give a voice to ordinary people without having to appeal to the exhibitionism of feelings or to stick to the rules of the traditional media agenda.

The fourth chapter of this book moves onto the study of TV consumption, referring to the attitudes of the audiences and how the digitalization of our cultures is modifying the forms of TV consumption. Leif Kramp discusses the notion of archive that is fundamental to understanding the transition from old to new media not only under a quantitative point of view. As it has been demonstrated by a recent range of books, exhibitions and conferences – the best and the most interesting was "Atlas" organised last year by Georges Didi-Huberman at the Museum Reina Sofia of Madrid – the notion of archive is the core of a modern and western conception but also one of the most important conquests of the postmodern ways of communication. This is why digital innovation today, that was celebrated in the nineties as the last outcome of "real time", is much more recognized for the number and the variety of information that consumer can get through the digital archives such as YouTube or even more through the social networks as a kind of living archive.

In Chapter 5 Fausto Colombo and Andrea Cuman are concerned to avoid adopting technological determinism in their analysis preferring to underline the nature of television as a cultural device. One of the most important innovations made possible by digital media, is the recovery of the "myth of gratuitousness" that was much stronger at the beginning of the commercial TV era and has now been re-established by the digital free circulation and free downloading of TV content. The value of TV consumption at the time of Web 2.0 is generated by different dimensions: the individual and self-managed (as a sort of economy of attention) and the collective way of consumption that creates a sort of a "bottom-up or horizontal circuit" where the users of a social network can assign value to the content through peer networking.

The point of view of outsiders on American TV is explored in Chapter 6 by Eleonora Benecchi and Giuseppe Richeri with special attention to the consumption of TV series and the way they can activate phenomena of fandom. Their analysis begins from a quite inspiring issue: the consideration that it is not useful to invest in marketing and advertising if you do not know how to involve the community of fans that follows your series. Comparing the traditional literacy and traditional media fandom with the Web 2.0 approach, the authors underline an extension of values and practices that defines the complexity of the contemporary mediatic consumption. These range from the exhibitions of "creativity, critical approach, participation" to the process of "lauding, preserving, collecting, scrutinising and being passionate about a popular TV" that turns this social phenomenon into a kind of paradoxical mainstream subculture.

The seventh chapter is dedicated to points of contact between TV and other new technologies such as the newest generation of mobile phones and Internet in a dimension that could be termed "Glocal". The paper written by Juan Miguel Aguado, Claudio Feijóo, Inmaculada J. Martínez and Marta Roel starts from the dimension of mobile phone content consumption and tries to define the trajectories of interaction between this medium and other more general innovation such as the multi-screen convergent television. Here, the notion of ecosystem is a constitutive step in their analysis and shows the field of competition between two different technologies, considering their different positioning and possibility of an intersection between their respective services. Starting from a comparative approach between different geographic areas (USA, Europe and the Asia-Pacific) the authors show the nature of the limits to this convergence. It is basically economic, technological and normative, but, as the meaning of "mobility" anticipates, it also concerns the users and their cultural dimension.

Andrea Miconi's Chapter 8 aims to focus on the role played by micro-TV stations, during the chaotic transition between broadcasting system and new media system, as happened in the Italian cultural industry (1990-2010). This transition can be divided into two main periods: the first has been characterized by several amateurish broadcasters, such as "neighbourhood TV", deeply grounded in the culture and needs of local territories, while the second is ruled by the rise of Web-TV stations, showing a new technological asset and, to a great extent, a new way for the Italian mass media system as well.

The Ninth Chapter is recognition of the ecology as a consumer based approach in the two main dimensions of social and individual consumption. Jakob Bjur wants to verify the idea of a progressive and increasing collapse of the Television era through research based on a general mapping of the contemporary scenario. His analysis uses the Swedish People Meter data as a representative sample of the national television audience that gives static information about the channel chosen by the audience and also dynamic information about the audience activities. This allows distinction between individual TV consumption from the dyadic or collective consumption and shows how a TV becoming even more "social" is not just a mainstream process, but more an expression of a certain type of target with some specific characteristics.

Bartolomeo Sapio, Tomaz Turk, Stefano Livi, Michele Cornacchia, Enrico Nicolo', Filomena Papa investigate in Chapter 10 the Italian diffusion of the digital television (DTV) with particular attention to the influence of end-user variables in the adoption strategy based on the interactive payment service and its security issues. Digital TV is not just a new device and a new way of consumption but also a new experience that creates a different fidelity with the audience. The authors show the results of a field study on the Italian T-government project "Services for citizens via DTV" carried out on a sample of 300 users, selected in three main areas of the country (North, Centre and Rome, South).

As the reader will no doubt appreciate, the variety and multi-disciplinarity of the approaches presented in this book offer considerable added value to a deeper comprehension of the new TV scenario. From the emotional construction of devices and contents to the strictly technical design of functions and services, we nevertheless find the dialectic between individual and "social" consumption, massive diffusion and customisation, professional managing and neo-amateur uses. Beginning with McLuhan's original thoughts about the environments in which messages are conveyed, through these chapters we re-discover that the individual user is at once the core and the generator of the new communicative platforms that comprise the contemporary multi-channel and integrated mediatised ecosystem.

References

Bolter J. D., Grusin R. 1999. Re-mediation: Understanding New Media. Cambridge: MIT Press.

Castells M. 1996. The Information Age: Economy, Society and Culture. Vol. I: The Rise of the Network Society. Cambridge et al.: Blackwell.

Fortunati L. 2009. Old and New Media, Old Emotion. In: J. Vincent & L. Fortunati (eds.). Electronic Emotion: The Mediation of Emotion via Information and Communication Technologies. Oxford: Peter Lang.

Foucault M. 1978. The History of Sexuality. Vol. I: An Introduction, translated by Robert Hurley. New York: Pantheon.

McLuhan M. 1964. Understanding Media: The Extensions of Man. New York: Routledge.

McLuhan M., Nevitt B. 1972. Take Today: The Executive as Dropout. New York: Harcourt Brace Jovanovich.

McLuhan S., Staines D. (eds.). 2003. Marshall McLuhan: Understanding Me: Lectures and Interviews. Cambridge: MIT.

Morin M. 1962. L’esprit du temps. Essai sur la culture de masse. Paris: Grasset.

Part I

The Role of Emotion in the New Television Ecosystem

Leopoldina Fortunati & Sakari Taipale

Adoption of New Forms of Television and Emotion in Five European Countries

Introduction

Digitalisation of broadcasting techniques is transforming the landscape of television across the world. In Europe, the regulation and adoption of terrestrial television broadcasting in particular have received a lot of public attention within the last few years. Along with terrestrial television, Internet Protocol TV (IPTV) is considered to have remarkable market potential as broadband connections have become faster and easier to access (Thompson, 2007; Simpson and Greenfield, 2007). Despite the emerging forms of TV, a remarkable number of Europeans are still watching analogue terrestrial television or making use of cable and satellite technologies based on analogue or digital solutions.

It is our intention to look at the extent to which the populations of the five of the most populated and affluent European countries – Italy, France, Germany, the United Kingdom and Spain – have adopted these new forms of television. The first aim of the study is to explore how the television audience has split into several different audiences according to the various platforms now available and identify the socio-demographic structures of these different TV audiences. Therefore, although the history of television studies has developed largely around the question of the medium's influence (Silverstone, 1994, p. 132), in this paper we want to focus on viewers' identity by drawing portraits of the audiences of the new forms of television. Our first hypothesis is that the oft-discussed fragmentation and individualization of the paths of media consumption (see for example Bjur, 2009) does not impede the formation of different audiences that are shaped by certain shared characteristics. The second aim of this study is to focus on the broad field of the practices of use of television in terms of a particular aspect: the emotional fabric that people attach to television and its new forms. According to Lull (1980), social use of television in the home is of two primary types: structural (environmental and regulative) and relational (communication facilitation, affiliation/avoidance, social learning and competence/dominance). The emotional fabric crosses both types of social use in the sense that, for example, companionship and fun are related to environmental dimensions. Investigating it is particularly important since emotions coincide to structure reality, to outline a predictable social reality, and to provide models of elaboration of expectations (Flam, 1990; Vincent and Fortunati, 2009). Our second hypothesis is that the pluralisation of television forms has brought a restructuring of the emotional response that the TV audience has traditionally made to analogue television. Investigating emotion is important since it is an important part of social action, constitutes one of the three dimensions of the model of social action (which consists of norms, emotions and reason; Flam, 1990; Turnaturi, 1995; Fortunati, 2009), and in our specific case is a good way to capture indirectly the reaction to and reception of the different forms of television. Interrogating audiences is not currently the best strategy as we share Derrida and Stiegler's concern that TV viewers are all in a state of illiteracy in respect of the image (Derrida and Stiegler, 1996/1997, p. 64).

To tackle these questions we will analyse survey data collected from Italy, France, the UK, Germany and Spain (N=7255) in 2009. The survey, funded by Telecom Italia, partially replicates research carried out in 1996 on the same countries (N=6609) (Fortunati, 1998). With the help of these data we will not only be able to break down the profile of television viewers of the different forms of TV, but also to present multivariate models to predict the ownership of these new forms of television. We are primarily interested in studying the new, emerging, forms of television, but of course we will take the analogue terrestrial television as a comparator. Second, we are interested in investigating the emotional fabric that characterises people's relationship with the television and its different forms.

This chapter begins with an overview of previous literature on the role of television, and particularly on how TV viewing is connected to the emotional fabric of audiences, as well as to the debate on technological convergence/divergence. We will then address the issue of the new forms of television, considering how the spatial-temporal categories of television use are restructuring as a consequence of their adoption. The question of whether the new forms of television are changing the emotional dimension of TV viewing will be discussed in particular. After the review of previous literature, we will turn to our data and measures for illustrating them. Of the many possible ways of conceptualising and defining emotion, we have decided to operationalise them by applying James A. Russell's (1980) circumplex model of affect. This model has been tested over the last three decades and has proved to be appropriate for measuring emotions in various research settings (Martin et al., 2008, pp. 225-227). After presenting and discussing our analyses and results regarding the profiles of the audiences of the new forms of television and the emotional fabric applied, we will draw some final conclusions.

The New Forms of TV: A Divergent Process

Since its domestication television has had a pivotal role in structuring the use of time in the economy and ecology of everyday life of family members. Programmes such as daily news, soap operas or series and major sporting events set routines, rituals and rhythms for everyday life (Yoshimi, 2010, p. 552; Kortti,

2010, p. 9). The scheduling and execution of other chores, if they are not absolute necessities, have been subordinated to them (Gauntlett and Hill, 1999, p. 35) or have been carried out in juxtaposition, as previous research carried out in 1996 showed (Fortunati and Manganelli, 1998). Besides the importance of television in the spatial and temporal organization of family life, as James Lull observes (1980, p. 197), TV use also plays a strong role in setting conversational agendas, developing socialisation and interpersonal interaction patterns within families, influencing the use of languages, patterns of speech and thought. Here, however, we are specifically concerned with investigating how television conveys a wide spectrum of emotions which according to some studies are discussed and shared by viewers along with ideas, information and symbols (e.g. Huston et al., 1995; Harper, Regan and Rouncefield, 2006). Silverstone, for example, argues that TV may provide food for sustaining and managing conversations as well as feelings of dependence, security and attachment (Silverstone, 1994, p. 11, p. 40). On top of this, television has the capacity to create, strengthen and release the emotional tensions that can develop within families (Kortti, 2010, p. 5).

The new forms of TV talk against the heated debate about the convergence process where various technologies are seen merging together in contemporary societies (Jenkins, 2006). Fortunati (2008) has discussed this elsewhere and argued that divergence (the opposite process) is an equally important issue. Our object of investigation – the new forms of television – is proof of the relevance of the divergence process. This process of technological divergence, clearly embraced by users, has in its turn also had consequences for consumption behaviours and lifestyles, which have become more personalised and individualized. In fact, it is a common trait of all new forms of TV that they are able to provide viewers with a higher level of personification with regard to space, time and content of TV viewing (e.g. Moran, 2010, p. 293). Earlier studies report, for instance, that the users of mobile television and IPTV appreciate most watching programmes from archives whenever it fits into their own schedules (Södergård, 2003). The temporal personification of TV-watching means that people have the opportunity to break away from those collective rhythms of daily life which were built from shared TV-watching at home with other family members, and from those constraints deriving from the different time schedules, tastes and desires of family members.

The Internet has made it possible to watch television programmes online when it fits with one's own schedule, to become a "devoted" fan, to orientate and influence broadcasting in so many ways and to share these experiences with others in online communities (Kortti, 2010, p. 13; Wohn and Na, 2011). This is to say that a higher degree of individuality in TV-viewing behaviour does not automatically signal people's "alienation" because new forms of television can also be shared with others and be co-consumed (e.g. Harper, Regan and Rouncefield, 2006).

These studies actually indicate that the technological divergence of media may actually increase the communality of television.

At the same time, this divergent process has involved the emotional response experienced by different TV audiences to the new forms of TV. Many authors have recently investigated the emotion of TV audiences in order to understand how people orient themselves in front of new types of programme, especially reality shows that aim at the intimate engagement of the audience and the unleashing of their emotions (e.g. Aslama and Pantti, 2006; Gorton, 2009; Kavka, 2008; Ellis, 2009). The cross-cultural study that we discuss here, however, is not so common in audience studies.

In our study the divergent process is explored in the context of five EU member states: Italy, France, Germany, the United Kingdom and Spain. After the phase of "classical television", roughly from the 1960s to the 1980s, characterized by national analogue broadcasting companies and a small number of channels (Gripsrud, 2010, pp. 77-78), satellite and cable television were introduced in all these countries (Van der Broeck and Pierson, 2008). Gripsrud (2010, p. 80) writes that these two technologies were adopted very rapidly and that they were well received by consistent groups of audiences in Western Europe. He adds that the launch of cable and satellite technology took place in chorus with the success of neo-liberal deregulation media policy, and the success of these technologies was guaranteed by a demand for different sorts of channels. The diffusion of satellite and cable TV is far from being homogeneous in the countries considered in this study, however. For example, cable TV in Italy never really developed because it has always been opposed by the public RAI-TV service and by the government. There have been many attempts to implement this form of television, such as Stream (by Telecom Italia) and TV by Fastweb, but they have not succeeded. Satellite TV was embraced more enthusiastically in Germany and the UK than in the other three countries.

The newest technology studied in our article is IPTV, which is still in its infancy in all of the explored countries. According to Thompson's (2007) report, France had the largest number of IPTV subscribers in 2006, a trend confirmed also in successive years, according to the report "A Sampler of International Media and Communication Statistics 2010" (Leckner and Facht, 2010). Furthermore, Thompson's report also confirms that France, the UK and Spain were amongst the most advanced countries in Europe in terms of the switchover from analogue to digital terrestrial broadcasting. The EU member states have committed to switch off analogue terrestrial TV by the end of 2012 and it seems that this goal will be met by almost every country.

Data and Methods

This chapter is based on a broad survey that was carried out in Italy, France, the UK, Germany and Spain in 2009 (N=7255). The sample is representative of the populations of these countries and is structured as follows: Italy (N=1398), France (N=1424), Germany (N=1919), the United Kingdom (N=1411) and Spain (N=1103). The data were collected by means of a fixed telephone survey. In this study we used weighted data in order to correct distortions (related to age, education, ownership of a computer and access to the Internet), which affected the correct representation of the various quotas of the sample. This survey was an adjusted replication of the first survey carried out in 1996 in the same European countries and with representative samples of the related populations, and, when appropriate, we use those data for comparison with the new data collected.

The socio-demographic variables included in the analyses of the present article were gender, age, education, family typology, the size of the city residence, and country. Among the respondents, 3,551 were males (48.9%) and 3,704 were females (51.1%). The age of respondents was measured by years and, afterwards categorized into five groups (14-17, 18-24, 25-44, 45-64 and 65 years and over). The typology of families was divided into singles, couples without children, and couples with children, single-parent families and mixed families (all the other types of families). Respondents' main activity was broke down into five categories: employees, house persons, unemployed, retired, and students. Education level was divided into the following categories: low (primary and secondary school diploma), middle (high school diploma) and high (college/university degree or higher). Finally, seven categories were distinguished with respect to city size (cities with fewer than 5,000 inhabitants; 5,000-10,000; 10,000-30,000; 30,000-100,000; 100,000-250,000; 250,000-500,000; 500,000 or more). Other socio-demographic background variables such as macro-region, income and professional status (even if investigated in this study) were excluded after careful analysis.

As regards the method, the article deploys bivariate statistics with a set of related tools, such as Chi-square tests and standardized residuals[1], and a multivariate method called logistic regression analysis (LRA). The logistic regression analyses with an entered method were executed to tackle the first question, whereas bivariate methods were used to study both the first and the second research questions. With regard to LRA, the Hosmer and Lemeshow test was used to indicate goodness-of-fit. To find out the overall proportion of the variance explained by our LRA models, we referred to Nagelkerke statistics (Tabachnick and Fidell, 2007,

1 The analysis of standardized residuals is based on the identification of the cells of a contingency table, which are responsible for a significant overall chi-square. Values outside +/- 1.96 are interpreted as statistically significant (e.g. Everett, 1992, pp. 46-48; Field, 2009, pp. 698-700). To simplify the analysis, however, we read only the positive residuals.

pp. 459-461). A dichotomous question about the possession of each of the various forms of TV was used as a dependent variable in the regression analysis, and a set of nominal, ordinal and dichotomous variables measuring demographic and spatial factors (Tabachnick and Fidell, 2007, p. 437) were entered as independent variables. Additionally, the potentially harmful impacts of outliers were ruled out.

Measures

Television. To study the new forms of television we used the question: "Does your household subscribe to any pay-TV channels?" Analogue television users were operationalised by excluding the pay-TV channel subscribers from those respondents who possessed a television. Respondents were asked to choose between multiple choices, which were Satellite TV, Digital Terrestrial TV, IPTV (receiving TV via Internet), and Cable TV. Additionally, respondents were given a chance to answer "No" or "Don't know/Can't remember". The question did not distinguish between analogue and digital technology with respect to Satellite and Cable TV.

Emotions. To find out the feelings people associate with television we used the same battery of emotions that was in the 1996[2] survey. Then we applied Russell's circumplex model to reduce the emotion scale. The model sees emotion as organised according to a circular structure ("circumplex"), which is a two-dimensional space consisting of pleasure-displeasure and arousal-sleepiness (or high-low arousal) axes (e.g. Russell, 1980; Russell, Lewicka and Niit, 1989). The final emotion measure is made up of four emotional categories: excitement (pleasure/arousal), distress (arousal, unpleasant), depression (unpleasant/sleepiness) and contentment (sleepiness/pleasure). Each emotional category represents one quadrant of the circumplex model of affect.

2 In the 1996 survey a pretest consisting of an open question was applied. Respondents asked to answer spontaneously the question "What emotions or feelings do you have about each of the following means of communication? Please just give one word off the top of your head for each one". Answers were classified by the researchers, who closed down the question in the questionnaire into 20 predetermined categories: interest, enthusiasm, curiosity, anxiety, irritation/annoyance, joy/pleasure, satisfaction, frustration, anger, embarrassment, surprise, relaxation, companionship, fun/happiness, indifference, boredom, others, nothing in particular, don't know, no response.

Results

Portraits of the Audiences of the Various Types of Television

In this chapter we begin by depicting the portraits of the audiences of the various forms of television. As a first step we present bivariate relationships between various types of television and structural variables in Table 1. The data reveal that women more typically use analogue terrestrial television than men, although analogue television is still the most common form of television, among both genders. Conversely, men own and use cable and satellite television more widely than women. Regarding the age of respondents, our analysis unsurprisingly shows that analogue terrestrial television is most widely used among the oldest users (65+), whereas the youngest respondents (14-24-year-olds) show the highest proportions when satellite TV or IPTV is considered. Respondents aged from 35 to 44 present the highest proportion of digital terrestrial television. Regarding cable television, our data show no major age-based differences.

In addition, the data show that level of education is strongly associated with the access to or possession of IPTV and satellite television. Highly educated respondents are more likely to possess these types of television than respondents with a medium and, especially, a low level of education. Almost 60% of respondents with a low education level are still users of analogue television in the countries studied. When looking at respondents' main activities, we can see that students stand out from other groups, being more typically users of satellite and Internet television. Respondents in employee positions are forerunners in possessing digital television, and retired people report higher levels of analogue television possession than other groups.

Family composition reflects differences in the possession of various types of television as well. Single respondents are less likely to have satellite and digital terrestrial television than respondents living in other types of families. Couples with children, on the other hand, most typically have the latest television technology in their homes. In terms of satellite, digital television and IPTV possession, they are slightly ahead of other family types.

Lastly, we also explored how two spatial factors, the size of the city of residence and country, are related to various forms of television. First, as one may expect, the newest forms of television, such as digital terrestrial television and IPTV, but also satellite TV, are typical especially for those respondents who reside in large cities of 500,000 or more inhabitants. Correspondingly, these respondents are less likely to have analogue television than other groups. Compared with smaller towns, IPTV and satellite television are also very common in the cities with 250,000 to 500,000 inhabitants. In addition, it is interesting to note that satellite television is rather common in the smallest villages (fewer than 5,000 inhabitants) and small towns (10,000 to 30,000 inhabitants). This is partially explained by the poor availability of wired service, such as the cable and broadband needed for IPTV, in the smallest villages. Second, the countries studied are very different from one another in terms of the kind of television services they provide to their citizens. In Italy, cable television service is basically non-existent, but satellite television is more common than in other countries. Additionally, Italy still has a rather high proportion of analogue terrestrial television users. Digital terrestrial television is used more widely in France (15.5%) than in other countries. France is also the country with the highest, albeit still very low, number of IPTV users (3.0%). More than one-third of German respondents have subscribed to cable television, but in Germany the proportion of digital television users is still rather low. The UK sample shows a relatively high proportion of digital terrestrial TV users. The UK also has the second highest proportion of IPTV users in our data set. Spanish respondents, however, registered the biggest presence of analogue terrestrial television, and the lowest figure for satellite television.

	Satellite TV	Digital Terrestrial TV	IPTV	Cable TV	Analogue Terrestrial TV	Total	N	Chi-squares, sig
Gender								
Male	17.1*	12.3	1.5	19.3*	49.7*	100.0	3423	51.866***
Female	13.8*	10.7	1.4	16.0*	58.2*	100.0	3614	
Age								
14-24	18.5*	13.0	2.6*	17.1	48.7*	100.0	1069	137.479***
25-34	17.6	13.0	2.5*	17.9	48.9*	100.0	1060	
35-44	15.2	14.1*	1.5	18.0	51.2	100.0	1222	
45-54	15.1	12.6	1.6*	16.2	54.5	100.0	1198	
55-64	13.0	11.1	0.9	17.9	57.0	100.0	966	
65+	13.5	6.6*	0.0*	18.3	61.6*	100.0	1523	
Education level								
Low education	12.7*	8.7*	0.5*	18.9	59.2*	100.0	2915	130.392***
Middle education	17.0*	14.0*	2.0*	16.5	50.5*	100.0	2656	
High education	18.8*	13.2	2.4*	17.7	47.9*	100.0	1316	
Activity								
Employee	15.5	14.5*	1.8	18.0	50.3*	100.0	3682	167.509***
House person	13.6	9.0	2.4	17.9	57.1	100.0	587	
Unemployed	18.1	9.2	1.8	20.3	50.6	100.0	271	
Student	22.3*	11.6	2.7*	12.6*	50.8	100.0	524	
Family composition								
Single	9.4*	9.1*	1.1	21.4*	59.0*	100.0	1752	180.825***
Couple, no children	14.8	10.4	0.9	20.1*	53.8	100.0	1688	
Couple with children	20.1*	14.2*	2.1*	14.1*	49.6*	100.0	2508	
One parent with children	16.0	8.0*	1.5	13.7	60.8	100.0	401	
Blended families	15.6	12.5	1.1	18.0	52.8	100.0	646	

	Satellite TV	Digital Terrestrial TV	IPTV	Cable TV	Analogue Terrestrial TV	Total	N	Chi-squares, sig
City size (inhabitants)								
Fewer than 5.000	24.2*	9.9	0.5*	11.9*	53.5	100.0	628	146.097***
5.000 - 10.000	19.3	11.5	0.7	15.6	52.8	100.0	538	
10.000 - 30.000	21.7*	10.9	0.5*	15.6	51.3	100.0	797	
30.000 - 100.000	15.9	11.1	0.9	17.1	55.0	100.0	976	
100.000 - 250.000	14.5	10.2	2.3	17.5	55.5	100.0	794	
250.000 - 500.000	10.5*	14.7	3.4*	13.9	57.5	100.0	619	
500.000+	14.6*	14.5*	2.3*	20.8*	47.7*	100.0	1474	
Country								
Italy	22.5*	13.0	0.6*	-	64.0*	100.0	1343	1097.258***
France	18.6*	15.5*	3.0*	13.3*	49.6*	100.0	1378	
Germany	18.1*	6.4*	0.4*	36.5*	38.4*	100.0	1839	
United Kingdom	8.1*	15.4*	2.6*	15.3*	58.7*	100.0	1398	
Spain	7.4*	7.9*	0.7	15.7	68.3*	100.0	1079	

Sig.=*=p<0.05, **= p<0.01, ***=p<0.001.
* Standardized residuals higher than 2.0 and lower than -2.0 are statistically significant

Table 1: Type of television by personal and structural variables

What Predicts the Adoption of the New Forms of Television?

As we have seen so far the adoption of new forms of television is dependent on various personal and structural factors. But what are the main predictors for this adoption? To answer this question we present the results of logistic regression analyses (LRAs), which were carried out separately for each type of television in order to depict the profiles of their audiences (Table 2). LRAs were implemented with an entry method so that only the predictors that significantly increased the capacity of a given model to classify the observed data correctly were included in the final models. As the adoption of new forms of television is highly dependent on national policies and available infrastructures (see also Table 1), a country variable is included in the models to level out the differences that would otherwise be mirrored by other variables.

The predictive power of the models, measured by Nagelkerke's R2, ranged between 0.081 and 0.258 (from 8 to 26%) The Hosmer-Lemeshow tests signalled a good fit for the new types of television, as values ranged between 0.124 and 0.942. Given the very broad diffusion of the traditional television, it was not surprising that the fit was not as good for the analogue terrestrial television.

Satellite television. LRA shows that men are more likely to possess a satellite television than women, and highly educated people are almost 1.5 times as likely to own satellite television as people with a low level of education. Family composition is also significantly related to the possession of satellite television: compared with single respondents, respondents living in a family made up of couples with children and in blended families are twice as likely to have a satellite television; there is also a statistically significant difference in the same direction between singles and couples without children. When looking at two spatial factors, city size and country, our results show that respondents living in small villages are more likely to subscribe to satellite television services than their urban counterparts. LRA also confirms that satellite television is more characteristic of French respondents than of the others.

Digital Terrestrial Television. Gender and age are not statistically significant predictors for the possession of digital television. Our model, however, shows that compared with respondents with a low level of education, those with medium and, especially, those with high levels of education more typically have a digital television subscription. We also found that compared with the respondents who are actively participating in working life only retired people were less likely to have this type of television. Between employees and other activity groups (i.e. house persons, unemployed, and students) we were not able to detect any differences. When it comes to family composition, the model indicates only small differences. Compared with a reference group, i.e. singles, only respondents living in a family made up of couples with children were slightly more likely to have a digital television. City size is associated with the possession of this television type as well, although the variation between cities of different size is relatively small. Compared with the respondents of the smallest villages (fewer than 5,000 inhabitants), who have the lowest level of adoption, the respondents of the largest cities (more than 500,000) are about 1.6 times as likely to have a digital television. Finally, the country of residence was tightly associated with adoption. Compared with the reference country, France, the UK respondents especially, but also those of other nationalities, use digital television less often. This is a sign of the asynchronous adoption of digital television in Europe.

Internet Protocol TV. IPTV is still a novel form of television broadcasting in Europe. Owing to the small number of respondents, we found only two powerful predictors for its adoption. First, and predictably, the country of residence is strongly linked to its adoption, owing to differences in the availability of IPTV services in different countries. French respondents are far ahead of all other respondents in the use of IPTV, although the adoption rate of IPTV in general is still very low in France. The other predictor that remained significant is the level of education: respondents with a middle level of education are 3.5 times as likely to possess IPTV and highly educated people are no less than six times more likely to have an IPTV than respondents with a low level of education.

Cable television. Like satellite television, cable television is more popular with men than women. Table 2 shows that subscription to cable television is more typical of lowly educated than highly educated people. It is also a characteristic of cable television that its users are typically house persons. Regarding family composition, our model shows that single people are the most typical group of cable TV users. The difference regarding respondents living in families composed of couples with and without children, as well as one-parent families, is statistically significant, whereas there is no difference between singles and blended families. Cable television is also more widespread in large cities than in smaller towns and villages, a fact, which is clearly related to the high building costs of wired infrastructure in remote regions. German respondents are almost five times and Spaniards 1.5 times as likely to possess a cable television compared with the French, who serve as the reference group. In addition, there is no difference between France and the UK. Italy did not provide cable television at the time of survey collection.

Analogue terrestrial television. The LRA model for analogue terrestrial television does not reveal major differences (although some of the differences are statistically significant) between men and women, education levels, activity categories or city sizes. People aged 45 or more are somewhat more typical owners of analogue television than younger age groups. Compared with singles, couples with children and mixed families are less frequent users of analogue television. The reason for this is that they have already started to use satellite television and digital television. LRA also shows that in those countries where newer forms of television are less common (especially Spain and Italy), there are still more users of analogue terrestrial television.

Predictors (reference groups in brackets)	Satellite TV	Digital Terrestrial TV	IPTV	Cable TV	Analogue Terrestrial TV
Gender (Male=0.Female=1)	0.755***	-	-	0.752***	1.145***
Age (14-24)	-	-	-		1***
25-34	-	-	-		0.941
35-44	-	-	-		1.134
45-54	-	-	-		1.497***
55-64	-	-	-		1.361*
65+	-	-	-		1.488**
Education level (low educ.)	1***	1***	1***	1***	1*
Middle education	1.19*	1.388***	3.520***	0.949	0.844*
High education	1.474***	1.770***	6.175***	0.657***	0.835*
Activity (Employee)	-	1***	-	1**	1*
House person	-	1.082	-	1.653***	0.741**
Unemployed	-	0.686	-	1.109	1.155
Retired	-	0.495***	-	0.964	1.117
Student	-	0.880	-	0.849	1.020
Family composition (single)	1***	1**	-	1**	1***
Couple, no children	1.634***	1.021	-	0.740**	0.902
Couple with children	2.041***	1.313*	-	0.749**	0.691***
One parent with children	1.368	0.732	-	0.575**	1.168
Blended families	2.042***	1.334	-	0.918	0.654***
City size (fewer than 5.000)	1***	1**	-	1***	1***
5.000 - 10.000	0.881	1.501*	-	1.402	0.919
10.000 - 30.000	1.102	1.317	-	1.362	0.864
30.000 - 100.000	0.855	1.457*	-	1.628**	0.929
100.000 - 250.000	0.702**	1.040	-	1.662**	0.968
250.000 - 500.000	0.613***	1.395	-	1.574**	0.967
500.000+	0.750*	1.660**	-	2.643***	0.678***
Country (France)	1***	1***	1***	1***	1***
Italy	0.519***	0.567***	0.176***	0.0003	1.970***
Germany	0.476***	0.399***	0.151***	4.777***	0.613***
United Kingdom	0.470***	1.421**	1.165	0.920	1.585***
Spain	0.199***	0.415***	0.269***	1.571***	2.209***
N	5599	5573	6882	5573	5572
Nagelkerke R2	0.081	0.09	0.10	0.258	0.109
Hosmer and Lemeshow Test (sig)	0.327	0.942	0.342	0.124	n.s.

Sig.=*=p<0.05 **p<0.01 ***p<0.001

Note. Only variables that significantly increased the capacity of the model to classify observed counts correctly were included.

Table 2: Logistic regression models for the adoption of various television types (Exp(B))

---

3 No cable television services provided in Italy.

Emotion with Regard to Television and its Various Forms

Up to now we have depicted the socio-demographic characteristics of the various audiences, providing a multivariate analysis that enabled us to predict the main variables responsible for the adoption of the different types of television. With regard to our research objectives the emotional fabric that these audiences experience in respect of television and its new forms remains to be explored. The emotion variable was not included in the same multivariate analyses with personal and structural variables as its capacity to distinguish the audiences of different television platforms was not as promising.

	Excitement		Distress		Depression		Contentment		Total	
	N	%	N	%	N	%	N	%	N	%
Enthusiasm	198	6.4							198	6.4
Interest	122	26.4							122	26.4
Curiosity	443	14.4							443	14.4
Joy/Pleasure	464	15.0							464	15.0
Surprise	123	4.0							123	4.0
Fun/Happiness	271	8.8							271	8.8
Anxiety			60	13.8					60	13.8
Irritation/ Annoyance			199	45.7					199	45.7
Frustration			81	18.6					81	18.6
Anger			95	21.8					95	21.8
Indifference					497	78.8			497	78.8
Boredom					134	21.2			134	21.2
Relaxation							211	18.3	211	18.3
Companionship							944	81.7	944	81.7
Total	3086	100.0	435	100.0	631	100.0	1155	100.0	5307	100.0

Note. "I do not know", "Nothing in particular" and "NA" are not reported in the table.

Table 3: Frequencies of emotion with regard to television and the application of Russell's model

In Table 3 we first report the emotional fabric as regards television in 2009 by applying Russell's circumplex model. The dominant feelings of audiences towards television appear to be highly positive as they are made up of excitement and contentment to the tune of 58.1% and 21.8% respectively. On the whole, negative feelings are really in the minority and account only for 20.1% of the whole emotion scale. These results however seem to contradict the emotional model of consumption, according to which the pleasure people feel when they consume a certain product diminishes over time (Elster, 1989). Thus they prompted comparison with the data collected in 1996. In producing this comparison, we found that in these thirteen years excitement and distress increased whereas depression and contentment decreased, as shown in Table 4.

	Excitement		Distress		Depression		Contentment		Total	
	N	%	N	%	N	%	N	%	N	%
1996	2452	47.0	379	7.3	734	14.1	1650	31.6	5215	100.0
2009	3086	58.1	435	8.2	631	11.9	1155	21.8	5307	100.0

Table 4: Comparison of emotion as regards television according to Russell's model in 1996 and 2009.

Although the proportion between positive and negative emotions does not show relevant changes, the internal composition of emotions changes. This is probably related both to the change in the quality of programmes and to the scope that audiences attribute to the use of television as well as to their various styles of use. These results allow us to hope that the new forms of television will follow the trend of classical television in contradicting the emotional model of consumption.

Next, we compared emotions across countries in order to show the influence of culture on the general emotional fabric (Table 5). Excitement, which is clearly the most common emotional feeling associated with television in every country, is much less in France and Italy than in Spain, Germany and the UK.

	Italy	France	Germany	UK	Spain	Total
Excitement	47.6*	51.3*	63.9*	63.0*	64.6*	58.1
Distress	11.6*	6.6	5.9*	8.2	9.1	8.2
Depression	10.1	19.6*	12.1	9.5*	8.7*	11.9
Contentment	30.8*	22.6	18.2*	19.3	17.5*	21.8
Total	100.0	100.0	100.0	100.0	100.0	100.0
N	1144	930	1367	948	919	5308

Chi-square 200.145, df=12, p=0.0001.
*Standardized residuals higher than 2.0 and lower than -2.0 are statistically significant

Table 5: Emotions associated with TV by country in 2009 (%)

Feelings of distress are more commonly reported by Italian respondents (11.6%), whereas Germans (5.9%) feel less distressed than the other respondents. Among those who reveal feelings of depression, the proportion of French respondents is rather high (19.6%) and it is almost double if compared with Spain (8.7%) and the UK (9.5%). Finally, among those who declare feelings of contentment Italian respondents greatly outnumber the respondents of other nationalities, especially Germans and Spaniards.

If we compare these figures with those derived from the 1996 data (Table 6), the feeling of excitement has increased in all the countries considered, except Italy, where it shows a clear decrease. In these thirteen years distress has grown in Italy and in the UK, and more moderately in Spain and Germany, whereas in France it shows an obvious contraction. Feelings of depression have decreased everywhere except France, where they grew. The emotion of contentment decreased in every country, however, with the sole exception of Italy, which has registered an increase in it. This means that whereas Germany, Spain and the UK have a common trend, Italy and France prove to be the exceptions. In Italy the excitement decreases in favour of the increase of contentment, meaning that Italian audiences have probably changed their attitude towards television, which has become less relational and more instrumental (use of the television as an aid to sleep or a relaxing/companionable tool). In France, where feelings of distress and depression have increased, it seems that audiences show more evident signs of discontent with television in the period considered.

	Italy	France	Germany	The UK	Spain	Total
Excitement	50.6	47.3	40.2*	57.9*	51.4	48.6
Distress	7.7	10.1*	5.1*	5.2*	7.0	7.0
Depression	15.3	12.8	12.7	13.0	14.7	13.7
Contentment	26.5*	29.8	42.8*	23.8*	26.9	30.7
Total	100.0	100.0	100.0	100.0	100.0	100.0
N	1144	930	1367	948	919	5308

Chi-square 150.079, df=12, p=0.0001.
*Standardized residuals higher than 2.0 and lower than -2.0 are statistically significant

Table 6: Emotions associated with TV by country (%) in 1996

To answer our second research question, we now analyse emotion regarding television by distinguishing its new forms. Table 7 reveals that when different forms of television are compared, no major differences in emotional associations emerge from the data, except those regarding the feelings of contentment.

It seems that the users of IPTV and digital terrestrial television are more content with their television than those who watch analogue television. Small differences remind us that different television technologies are to some extent mere distribution channels, and emotions are perhaps related more to content, services and the social context of viewing. In fact, more individualized services and content, as well as the possibility to interact with others, are also those properties that are expected of IPTV by its consumers (Shin, 2007). Finally, the table shows that the users of analogue terrestrial television report feelings of contentment less often than others. This may be an indication of their awareness of the better quality of digital broadcasting and the wider selection of related services which are not yet available to them or which they have not yet taken up.

	Satellite TV	Digital Terrestrial TV	IPTV	Cable TV	Analogue Terrestrial TV	Total
Excitement	58.9	56.0	52.2	57.9	59.0	58.3
Distress	6.9	7.8	4.3	6.7	8.8	7.9
Depression	10.3	9.6	8.7	11.7	12.8	11.7
Contentment	23.9	26.7*	34.8*	23.7	19.3*	22.0
Total	100.0	100.0	100.0	100.0	100.0	100.0
N	837	645	69	948	2461	5150

Chi-square 38,172, df=12, p=0.0001.
*Standardized residuals higher than 2.0 and lower than -2.0 are statistically significant

Table 7: Emotions associated with TV by the five types of television (%)

Discussion

Our first hypothesis regarding the feasibility of the portraits of the new TV poly-audience was confirmed by our data. In fact, by means of bivariate and multivariate analyses we were able to depict the structure of the socio-demographic characteristics of the audiences of the new forms of television and also the main predictors of their adoption. Our second hypothesis on the restructuring of the emotional fabric of audiences regarding television also received a positive answer from our data, although with some limitations. In fact our data showed that there are significant differences only regarding the feelings of contentment and only regarding digital terrestrial, IPTV and analogue television. It is likely that the novelty of these new forms of television has not yet allowed complete restructuring of the emotional fabric regarding the various forms of TV. Second, the high feeling of contentment with digital terrestrial television and the IPTV may relate to better quality of digital broadcasting and to the opportunity to choose the time and place of TV viewing especially when using IPTV. The small number of respondents with IPTV, however, requires us to be careful when interpreting the results. Furthermore, it is an interesting result that the emotional model of consumption does not work in the case of television.

Clearly, however, a virtuous circle with qualitative research is necessary for appropriate interpretation of the influence of the cultural variable on the emotional fabric regarding television and its new forms. Why is excitement, the most common emotional feeling associated with television in every country, much less widespread in France and Italy than in the other countries? Why do Italians report more feelings of distress and contentment? Why do the French report more feelings of depression?

Variations in the content of programmes and TV commercials watched in these five countries are likely to be reflected in the differences detected. For example, some studies argue that the UK viewers look for joy and humour from television advertisements, whereas the French audience prefers to watch more artistic and dreamlike advertisements (Whitelock and Rey, 1998). A comparison of television viewer types (Espe and Seiwert, 1986) also supports the idea that Germans and Britons, who are more interested in sport programmes (see also Livingstone and Bovill, 2001, p. 149), associate more feelings of excitement with television.

A study by Hargreaves and Mahdjoub (1997) provides one, but definitely not the only, explanation for the relatively high proportion of depressed audiences in France. When ethnic minorities in France were interviewed, many interviewees of the second generation reported being indifferent towards the acquisition of satellite television, which is particularly common in France (see also Table 2), both because they expected that their parents would control the selection of channels and because of their poorer command of their home country's language. A further explanation can be found from the fact that France respondents were the most disappointed in the worsening of the quality of TV programmes included in the 1996 survey (Fortunati and Manganelli, 1998, p. 165).

This study of the five European countries has highlighted some interesting differences between the countries as well between the surveys of 1996 and 2009 that would benefit from further study. In particular, future research should look, first of all, at television use in a more complex way by investigating programmes, content, and meanings. More generally, the dimensions and practices of the use of television should also be explored in a broader context, by considering time of consumption, contexts and modalities of use (personal, collective, familial). Second, the study should be situated in the broad ecology of media socio-technical systems in everyday life, analysing how the division of labour among them works and how they influence each other, intertwining and resonating.

References

Aslama M., Pantti M. 2006. Talking Alone: Reality TV, Emotions and Authenticity. European Journal of Cultural Studies, 9 (2), pp. 167-184.

Bjur J. 2009. Transforming Audiences: Patterns of Individualization in Television Viewing. University of Gothenburg: Department of Journalism, Media and Communication.

Derrida J., Stiegler B. 1996/1997. Échographies de la télévision. Paris: Éditions Galilée. (Italian Translation: Ecografie della televisione. Milan: Raffaello Cortina).

Ellis J. 2009. The Performance on Television of Sincerely Felt Emotion. The Annals of the American Academy of Political and Social Science, 625 (1), pp. 103-115.

Elster J. 1989. Nuts and Bolts for the Social Sciences. Cambridge: Cambridge University Press.

Espe H., Seiwert M. 1986. European Television-viewer Types: A Six-nation Classification by Programme Interests. European Journal of Communication, 1 (3), pp. 301-325.

Everett B. E. 1992. The Analysis of Contingency Tables (2nd ed.). Boca Raton FL: Chapman & Hall/CRR.

Field A. 2009. Discovering Statistics Using SPSS (3rd ed.). London: Sage.

Flam H. Emotional Man. I. The Emotional 'Man' and the Problem of Collective Action, International Sociology 5 (1990), pp. 39-56.

Fortunati L. (ed.). 1998. Telecomunicando in Europa. Milano: Angeli.

Fortunati L. 2008. Mobile Convergence. In: K. Nyiri (ed.). Integration and Ubiquity: Towards a Philosophy of Telecommunications Convergence. Wien: Passagen Verlag, pp. 221-228.

Fortunati L. 2009. Old and New Media, Old Emotion. In: J. Vincent & L. Fortunati (eds.). Electronic Emotion: The Mediation of Emotion via Information and Communication Technologies. Oxford: Peter Lang, pp. 35-62.

Fortunati L., Manganelli A. M. 1998. La comunicazione tecnologica: Comportamenti, opinioni ed emozioni degli Europei. In: L. Fortunati (ed.). Telecomunicando in Europa. Milan: Angeli, pp. 125-194.

Gauntlett D., Hill A. 1999. TV Living: Television, Culture and Everyday Life. London: Routledge.

Gorton K. 2009. Media Audiences: Television, Meaning and Emotion. Edinburgh: Edinburgh University Press.

Gripsrud J. 2010. 50 years of European Television. An Essay. In: J. Gripsrud & L. Weibull (eds.). Media, Markets and Public Spheres: European Media at the Crossroads. Bristol: Intellects, pp. 71-93.

Hargreaves A. G., Mahdjoub D. 1997. Satellite Television Viewing among Ethnic Minorities in France. European Journal of Communication, 12 (4), pp. 459-477.

Harper R., Regan T., Rouncefield M. 2006. Taking Hold of TV: Learning from the Literature. Proceedings of the 18th Australia Conference on Computer-Human Interaction: Design: Activities, Artefacts and Environments. New York: ACM.

Huston A. C., Wright J. C., Alvarez M., Truglio R., Fitch M., Piemyat S. 1995. Perceived Television Reality and Children's Emotional and Cognitive Responses to its Social Content. Journal of Applied Developmental Psychology, 16 (2), pp. 231-251.

Jenkins H. 2006. Convergence Culture: Where Old and New Media Collide. New York: New York University Press.

Kavka M. 2008. Reality Television, Affect and Intimacy: Reality Matters. Basingstoke: Palgrave Macmillan.
Kortti J. 2010. Multidimensional Social History of Television: Social Uses of Finnish Television from the 1950s to the 2000s. Television & New Media (online pre-print).
Leckner S., Facht U. 2010. A Sampler of International Media and Communication Statistics 2010, Nordic Media Trends, 12. NORDICOM. Gothenburg: University of Gothenburg. URL: http://www.nordicom.gu.se/common/publ_pdf/NMT.pdf (accessed 9 July 2011).
Livingstone S., Bovill M. 2001. Children and Their Changing Media Environment: A European Comparative Study. Mahwah, NJ: Lawrence Erlbaum.
Lull J. 1980. The Social Issues of Television. Human Communication Research, 6 (3), pp. 197-209.
Martin D., O'Neill M., Palmer A. 2008. The Role of Emotion in Explaining Consumer Satisfaction and Future Behavioural Intention. Journal of Service Marketing, 22 (3), pp. 224-236.
Moran A. 2010. Configurations of New Television Landscape. In: J. Wasko (ed.). A Companion to Television. Oxford: Wiley-Blackwell, pp. 291-307.
Russell J. A. 1980. A Circumplex Model of Affect. Journal of Personality and Social Psychology, 39 (6), pp. 1161–1178.
Russell J. A., Lewicka M., Niit T. 1989. A Cross-cultural Study of a Circumplex Model of Affect. Journal of Personality and Social Psychology, 57 (5), pp. 848-856.
Shin S. H. 2007. Potential User Factors Driving Adoption of IPTV: What Are Customers Expecting From IPTV? Technological Forecasting and Social Change, 74 (8), pp. 1446-1464.
Silverstone R. 1994. Television and Everyday Life. London: Routledge.
Simpson W., Greenfield H. A. 2007. IPTV and Internet Video: New Markets in Television Broadcasting. Burlington, MA: Focal Press.
Södergård C. 2003. Mobile Television: Technology and User Experiences. Report on the Mobile-TV Project. Helsinki: VTT.
Tabachnick B., Fidell L. 2007. Using Multivariate Statistics (5th ed.). Boston, MA: Pearson.
Thompson S. 2007. IPTV – Market, Regulatory Trends and Policy Options in Europe. Driving the Future of IPTV. Geneva: ITU. URL: http://www.itu.int/osg/spu/stn/digitalcontent/4.9.pdf (accessed 2 August 2011).
Turnaturi G. (ed.). 1995. La sociologia delle emozioni. Milan: Anabasi.
Van der Broeck W., Pierson J. (eds.). 2008. Digital Television in Europe. Brussels: VUB Press.
Vincent J., Fortunati L. (eds.). 2009. Electronic Emotion: The Mediation of Emotion via Information and Communication Technologies. Oxford: Peter Lang.
Whitelock J., Rey J.-C. 1998. Cross-cultural Advertising in Europe. An Empirical Survey of Television Advertising in France and the UK. International Marketing Review, 15 (4), pp. 257-276.

Wohn D. Y., Na E. K. 2011. Tweeting about TV: Sharing Television Viewing Experiences via Social Media Message Streams. First Monday, 16 (3-7). URL: http://firstmonday.org/htbin/cgiwrap/bin/ojs/index.php/fm/article/view/3368/2779 (accessed 2 August 2011).

Yoshimi S. 2010. Japanese Television: Early Development and Research. In: J. Wasko (ed.). A Companion to Television. Oxford: Wiley-Blackwell, pp. 540-557.

Nello Barile

The Age of Personal Web TVs. A Cultural Analysis of the Convergence between Web 2.0, Branding and Everyday Life

Introduction

This chapter aims to untie the different dimensions behind the strategy of self-broadcasting which is constitutive of the contemporary personal Web TV. Television is probably the main driver for the development of mass society and its decline overlaps with the end of the industrial society. The broadcasting communication and its main technologies are probably going to disappear, giving up their place to a new era in which media are not as functionally diversified as before. This approach is exemplified in the idea in the discourses on digital convergence, in the "eleputing" era (Gilder, 2000) or also the well-known remediation as "the representation of one medium in another" (Bolter and Gruisin, 1999, p. 45). According to the authors we can see how

> no medium today, and certainly no single media event, seems to do its cultural work in isolation from other social and economic forces,

while their main function is the way in which they

> refashion older media and the ways in which older media refashion themselves to answer the challenges of new media (p. 15).

If it is true that we are living in a new form of culture which is oriented and shaped by the power of software, it is no longer useful to analyse media as system of single communicative tools. In the dynamic of TV going "social", there is more than a remediation from old to new technologies, while there is rather an enlargement of uses and also a consumer driven process. If the digital is a new principle of content production, design and sharing, we cannot consider TV as just a medium. As Manovich (2008) underlined, re-evaluating Kay's and Goldberg's (1977) main idea, the huge software revolution pushes the computers evolution from the media to the state of "metamedium":

> the property that is the most important from the point of view of media history is that computer metamedium is simultaneously a set of different media and a system for generating new media tools and new types of media. In other words, a computer can be used to create new tools for working in the media it already provides as well as to develop new not-yet-invented media (Manovich, 2008, p. 68).

Metamedium is not just a container of different media, rather it is a different principle of relating data with contents. In fact it is possible to manage the same object with different applications in the way it can produce different effects as we change the point of view or the type of software manipulating data. More than a common convergence of different media we find a new principle of organization or a new technology which breaks from the traditional mediascapes and opens up the external environment as in the cases of ubiquitous computing, augmented reality or even some recent application of the hologram advertising. But I could even limit my analysis to the widespread technology of the smartphones.

This blurring of boundaries between different media is causing the traditional categories to vanish. In addition the nature of relationships between producer and consumer, the market positioning and the brand strategies are being deeply modified. It is not the case that the television marketing is less developed than the other more mature types of marketing (Mattiacci and Militi, 2011). It is not that the process of convergence is properly technological but rather it is more culturally driven by the consumer (Jenkins, 2006); new forms of television will be even more shaped by the consumer culture as well.

Those innovations are witnessing a triple process of convergence in which: a) traditional media are empowering their brand identity to entertain a new deep relationship with consumers; b) mass market brands are replacing TV and other media as content providers; c) people are transforming themselves in media – or better in a self-mass communication media – investing first in the transformation of their own identity in a brand.

Aim and Method

In the first part of this chapter I aim to analyse the direct integration between the Web 2.0 development and a new emotional capitalism. In the second part I will focus on a few technological and experimental projects which are trying to evolve and hybridise the conventional TV production, using the tools of blogging, social networking and User Generated Contents. In the course of the third part I will present and discuss briefly three case studies, which are clearly diversified under the cultural, economical and technological point of view.

So, the first question could be: how it was possible that the most advanced innovations of the digital technologies was the pre-condition to the liberation of the inner emotional capital, shown in contents or fuel for new communication and marketing strategies?

My analysis in the third stage follows the model proposed by Abruzzese and Borrelli (2001) which shows the development of the Italian television: 1) the generalist television; 2) the neo-television; 3) the post-television. The first stage is guaranteed by a straight separation between the context of production and the context of consumption. This separation is guaranteed by an ethical approach as a point of balance between two main "ideologies": the catholic and the communist. This is why the most seen (because it is unique) public TV channel, inspires its production towards the ideal of a conjunction between entertainment, driven by business and education, and driven by the ideologies.

The neo-television was conceptualised for the first time by Eco (1985) and it is characterized by the new forms and styles of the private television in which the advertising plays a fundamental role. On the wave of the general judicial deregulation, which opened the Italian TV market to private players. This moment represented the start of Silvio Berlusconi's media empire. As the reader may understand this process does not consist just of a business operation while it is mostly the cultural product of Berlusconi's shaping of the Italian culture. A strategy based not on the "separation" of elites with the audience but more on a "conflicting imitation". The age of the neo-television lasted throughout the eighties and will also survive with the third turning point. The post-television era is not just a local phenomenon. It is determined by the incredible diffusion during the nineties of the World Wide Web and will increase its power with the evolution to the Web 2.0. Post-television means the twilight of the generalist media style of communication which disseminates the seeds of the new media consumption, made possible by the digital media. Under a certain point of view, the success of the reality show is the pre-figuration of the future development of social media such as YouTube or Facebook.

This shift is well-analysed as a sort of paradoxical heritage from TV to the contemporary computers. As Uricchio (2010) has underlined in a recent paper, the computer emphasizes some aspects of the fulfilment of TV usage pushing them to their extreme limits. As TV, the computer is also a closed device that captures completely the user's concentration and, at the same time, opens the user's experience to the external world.

> As television [...] continues its latest pas the deux with the networked computer, the direction of flow is changing. In out online world, we read and write; we download and uplink; we consume and 'cut and paste' and produce. YouTube, of course, emblematises this participatory turns [...]. Although mainstream television is flirting ever-more intently with user-generated content (America's Funniest Home Videos; BBC Video Diaries and citizen journalism forms such as CNN: I-Report) much YouTube content is predicated upon the viewer/user's re-appropriation and re-contextualisation of existing televisual and film material (Uricchio, 2010, p. 38).

YouTube can be viewed as a recent addition into what Mirzoeff (2000) calls "visual culture". One of the most representative innovations of the Web 2.0 is certainly the modification of the TV concept subsumed by the participative logic of social media which creates a repository of mainstream TV fragments and crypto-amateur self-productions. This new way to consider the audiovisual experience turns the traditional TV contents into a new outcome which is social, participative, and challenging. Even if the high majority of contents are extracted from the broadcast production, YouTube started a revolution of a different way to communicate via audiovisual contents.

A New Frame for Confessions: Web 2.0 and the Emotional Capitalism

The expression Web 2.0 usually brings to mind an array of soft technologies that make use of development in hard technologies (more powerful servers, fibre optics, broadband etc.) This has led to a decrease in the distance between the processes of virtualisation and our daily lives. Consequently, the rhetoric of so-called user generated content (UGC) not only restores the ideal of the craftsman as an individual who puts personal commitment into a skilled activity (Sennett, 2008), but at the same time, creates one's own nemesis – that is, a new kind of autocracy in which technique is no longer simply a tool, but actually redefines social identities through new lifestyles. We could say that Web 2.0 generates a new emotional regime in which the clear opposition of modern identities and power is not that bright anymore; better it defines the perimeter of a field where two opposite yet complementary movements are set in motion: first, that of the large corporations who use brands to captivate their consumers and manipulate their daily lives; secondly, that of people who adopt Web 2.0 to shift from the local to a global level, in accordance with the renowned YouTube slogan: "Broadcast Yourself".

In other words, according to de Certeau's (1984) conceptualisation on the asymmetry between strategy and tactic, we could represent a pendulum swinging from the strategic to the tactical level or indeed from a tactic becoming strategic and vice versa a strategy that needs even more to become tactical. Everywhere, the centrality of the relational, the communal or the experiential side is celebrated, and perceived as a tangible sign of an irrevocable cultural turning point. It is equally clear that the communicative system disseminated by Web 2.0 does not suppress uniqueness of the individual expression; quite the opposite, it multiplies and emphasizes this feature in every possible way.

This is why today it is possible to introduce the notion of "punk capitalism" (Mason, 2008) as a way to identify an economic system that cultivates a close link with dispersion, transgression or even deviance as necessary vehicles for creative innovation in material and financial economics. This process has a long history, and this is why we should analyse it from an archaeological point of view. Furthermore the instance of self-presentation is not merely a modern concept, indeed it is quite modern in the way in which our technologies make this kind of action global. This is also the way in which this new lifestyle becomes almost hegemonic in the sense that nobody can refuse anyone wishing to adopt a strategy of a global self-presentation. This is why we must refer to Michel Foucault's History of Sexuality (1978) in which he seeks to undermine the framework of theories of repression which, he states, are nothing more than unconscious submission to that very logic of domination that such theories challenge. The demonstration of this reversal hinges on the mechanism of confession, originally intended to declare and absolve guilt in theocratic regimes. In the era of democracy, confession has become an instrument of production of the truth of pleasure. According to Foucault the Western individual has become a "confessing animal".

> The confession is a ritual of discourse in which the speaking subject is also the subject of the statement and it is also a ritual of power manifested by the presence of another. [...] It is the bond between the one who speaks and what he is speaking about within the intimacy of discourse that warrants the integrity of the confession. [...] The discourse of truth takes effect finally however, from the one from whom it was wrested and not from the one who receives it (Foucault, 1978, pp. 61-62).

In this enlightening passage, Foucault identifies and accurately describes the essential features of a communicative/therapeutic model which survives to his era, over the twilight of television and general media to establish a complete and systematic presence in the Web 2.0 domain. He argues that first confession is a self-productive discourse which is, at the same time, driven by a general idea of the listener who requests the confession. Second, there is the idea that the listener says nothing which is apparently a contradiction with the first statement, but it underlines how the speaker is self-oriented to confess because s/he has interiorised the general rule of the confessing society. Finally there is the way in which this ritual modifies principally the speaker and not the listener in a way which is reminiscent of a self-help therapy.

This self-expression device also works in a mass-mediatic regime or in a new media dimension even if the weight of the listener who takes an active part in the interrogation should be stronger. In this regard, Eva Illouz (2007) investigates the border between the fading sunset of the TV Empire, e.g. The Oprah Winfrey Show, and the new practices of self-promotion of social media; from YouTube to Twitter, from the leading characters of the blogosphere to those of personal web TV. She identifies "emotional ontology" – long term processes that were largely found in the organizational practices of American companies as well as in the great collective movements like feminism and self-help, up to the politics of showmanship in the eighties. The new capitalistic approach supports the "idea that emotions can be detached from the subject for control and classification" (p. 36). In this way they become the fuel of a generalized exchange between subjects, media, corporations, etc. and the main resource to be developed in their self-promotional policies.

However, digital and social media are the means for complete implementation of an emotional ontology – in other words, the ability to attract substantial public attention and involve large numbers of people into one individual's personal reality, especially one of suffering. Indeed, Illouz unsuccessfully uses the theme of suffering as a lever to challenge and contradict Foucault's concept described above. She states that the rapport between power and pleasure is positioned in Foucault as a dominant trait of human activity, whereas, in reality the French philosopher considers that term to be synonymous with sexuality or libido. Careful reading shows that Foucault's notion of sexuality is not in opposition to pain, but may even subsume it. Moreover, the deviant practices that Foucault alludes to, such as masochism, sadism or sodomy, lie on the border between pleasure and pain. Instead, Illouz suggests that pain is more a synonym of "authentic experience" in contrast to the barren landscape of the politics of showmanship in the 80s, which was based on a holy alliance of artificiality and hedonism. We can say that both, pleasure and pain, are deep emotional dimensions which attract the interest of old and new media. Maybe the first one has a stronger role in the definition of an authentic experiences and this is why confessional programs are so focused on its representation.

Logic of confession is not just a media production, and its coherence with other cultural phenomena developing from the 1990s fall of barriers between the front and back stages of politics, to the new era of civic engagement of young people; from the rediscovery of one's roots and a sense of belonging according to a glocal perspective to the revaluation of emotional intelligence at work and in social relationships. As a result, some of the earlier polemics re-emerge, especially those that argue against the mechanisms of confession, which, according to Agamben (2006) "desubjectivizes" rather than subjectivizes those who make use of this mechanism.

Let me reiterate that new technology is not the only way to explore these strategies based on self-expression, but rather it is its precondition that is the central point here; it makes known and magnifies a discourse that could very well take place in its absence. In this era, when technological change has been acclaimed like never before, we are surrounded by victims (Eliacheff and Lariviere, 2007) where the suffering, whether imposed or self-inflicted, of major actors in public life – but also of those who play a minor role – becomes the centrepiece of an effective strategy for self-promotion.

Being a victim is a winning role in a world where strong emotional intensity has led to a new form of capital that can be used to acquire various social advantages. The role of victim could even be considered part of a strategy that seeks to define one's existential positioning in a world that is increasingly "fluid" and elusive. As long as the phenomenon of victimisation is limited to marginal segments of the population, it holds a certain eschatological value. However, when the victim is a public figure or a current leader, this role adds a further sense of identity without diminishing its initial advantage. The role of the victim used as a shield or as a source of legitimisation is part of human history, but the new victim seeks to break down any separation between the actor and the spectator. This is why we will see how it can be used as well from famous and common people to underline some specific aspects of their intimacy as in the case of the new personal web televisions.

The sharp discontinuity between previous conceptions of the victim and its new version in the contemporary regime of the spectacular, should be pointed out once again. The recent work of Mayer-Schönberger (2009) depicts technology as a mere "externality" in comparison with the sentimental value of memory, to refer to a process which has now reached a high point – the loss of a life of oblivion. In direct contrast to Baudrillard's notion of "disappearance" (1994), nowadays we find ourselves in the situation where it is patently impossible to forget or be forgotten. If this process began with the same narrative of media, web and geo-location systems, it has now reached unprecedented levels for recording and tracking our every day activity, so that a sort of "electronic imprint" of the user stays in the "perfect" memory of the computer for a very long time.

Such considerations, which call to mind the recurring ghost of the surveillance society (Lyon, 1994), necessarily shift our focus on the cultural sphere, towards what Foucault used to call self-surveillance. Once again, the problem is not so much the way that new technologies overlap with our daily lives, establishing protocols for feelings, moods and relational networks. What is more worrying is probably the extreme way that people react to the total elimination of the boundaries between public and private, in an attempt to restore an accepted emotional order that has become confusing.

In this empty space, the so-called gossip society attains legitimisation in contrast to previous forms of mere chatter and gossip because of a twofold impossibility: first, that of giving up the use of these cultural technologies and second, that of "cognitive adaptation" (Mayer-Schönberger, 2009), which seeks to normalize a moral reaction to the loss of privacy. It is not by chance that one of the prevalent behaviours associated with Web 2.0 networking is obsessively checking who has visited our profile. This is tantamount to saying that the unconditional possibility of the confession – which entails large numbers of users spending time to create their digital profiles with highly experiential inputs (photos of travel and friends, personal tastes, hobbies, information about relationships) – implies willingness to check up on and, perhaps penalize those who come to browse the same content that has been freely offered to the public.

The confession mechanism first triggers voyeurism and then punishes it as a consequence of new waves of moral panic. Thus the problem lies in the social use of a "cultural technology" (Abruzzese and Borrelli, 2001) that requires profound comingling between the technology and its user. It is not by chance that the writer Lee Siegel (2008) suggests a variant of the famous McLuhan's statement "the medium is the message" (1964), entitling a chapter of his book "me is the message", in an attempt to disprove the combined myths of self-expression and "packaged selves" made possible by new media. In particular, Siegel first refers to the close relationship between the use of Web 2.0, about new existential positioning strategies, and about ethics or lifestyle of the new "creative" bourgeoisie dressed up as bohemians (the famous "Bobos in Paradise" by David Brooks (2000). But Siegel is also part of the old critical mould when he reflects on the key words "choice and access", which extend the ability to act of individuals; they break down the barriers of time and space while, at the same time, leading to a new form of homogeneity (Siegel, 2008, pp. 66-67).

The aforementioned confession mechanism, the new forms of victimization and the phenomenon of "packaged selves" can be seen as three models of existential positioning. The network provides a means to shape and then give global visibility to an individual profile and, at the same time, circulate the packaged self into even the remotest corners of everyday life, with the same promotional logic of a global brand. In some cases, new media acts as the environment; in others, the means; in yet others, it acts as the content of a more complex lifestyle that interacts dynamically with the technology.

Indeed, contemporary lifestyles are strongly conditioned by a set of values that mostly derive from the three categories of experience, emotion and relationship. The power that is released by the combined use of these categories creates a new environment which imposes a new form of discrimination based on intangible forms of capital. If Bourdieu's version of "cultural capital" (1984) was still governed by acquired mental structures (habitus), which regulated inner coherence, the lifestyles that predominate the present are deliberated in order to manage and exploit contradiction. Therefore, we might think in terms of an existential positioning that takes advantage of the assets of emotional, experiential and relational capitalism in order to construct new profiles.

Against the backdrop of contemporary forms of consumption, we no longer find much opposition to "instrumental rationality", but rather a new form of "instrumental emotionalism" which exaggerates some aspects of the emotional capital. Perhaps for this reason, the most recent marketing theories have revisited some of the essential concepts of philosophical trends of the 20th century. The theoretical foundation of Schmitt (1999), father of experiential marketing explicitly refers to "Erlebnis" as an adaptation of phenomenology to marketing. From the former, Schmitt derives the notion that the world is not external to the individual; rather it is the field within which one produces thoughts and perceptions (Schmitt 1999, p. 60), but he also considers experience as an induced and not self-generated process. This is why brands turn themselves from Identity providers to experience providers, from a definer of identity (Brand Id.) through naming, logo, slogan awareness and image – to a symbolic environment which works through senses or emotions; through creative relationships and a range of lifestyles.

Schmitt himself emphasizes the close relationship between the emotional and experiential spheres, so that every experience is connected to an emotion and in most cases, the name of that emotion (such as hate, love, attraction, etc.) is used to describe the experience that produces it. The category of experience is thus considered to be all-encompassing, omnipresent and inescapable as well as able to cut across both virtual and real space. For this reason, new marketing approaches assign an unquestionably positive value to experience. Whether material, cognitive, affective, sexual, playful, sensory etc., every social group seeks to accumulate a significant number of experiences. It is not a coincidence that this highly communicative format is extended to everything that provides feelings through an experience; from television dramas to supermarket displays. Instead the world of emotions is usually activated by the use of several technologies (Vincent and Fortunati, 2009), in the world of social networks even more important are the relations between experience and emotions. There are several manifestations of these relations in the normal use of the social network but it appears more obvious in special moments or events. For example, if a person decided to leave Facebook, they would discover that the Facebook network, in attempt to prevent this choice, was ready with a "gallery of horrors" made up of all your friends' profiles, which paraded across the screen "ripping" their hair out and begging you not to go away – "Oh please stay with us"; "It was so great to be together"; "We could still have so much more fun" etc. – these are just a few of the phrases assigned to real persons who are quite unaware that their virtual identities have been appropriated, apparently saying things that they may not actually think. Of course, it is only a game, but it is also a demonstration of how the social network apparatus covertly makes use of our scheme of relationships to reinforce loyalty to its own advantage. This assertion would take us back to an ominous other-directed vision if it were not for the fact that our Web 2.0 connections are the projections or the empowerment of our real relations system and this is the very link that binds us to the network itself. In other words, we belong to the network because it is the formal representation of our real system of relationships, while we might turn away from the technological platform that allows us to manage them, we could never give up the relationships themselves even if they are given impetus by the network. In the most advanced stages of digital technology expansion, the social network has clearly shown that the real content of its operation is the user and the set of connections that he or she develops over time – in short, the relationship itself.

How TV Frame is Re-Incorporated in the Web 2.0 Logic

Today TV becoming "social" is a sort of natural development of the global broadcasters involving channels, programmes or just single personalities. One of those new experiences proposed to the audience is the program called "Fan It", which sounds like an implicit reference to the "fan culture" and promotes the dynamic integration between TV and social media.

> NBC on Thursday unveiled a new program called 'Fan It' that will offer prizes to fans of its shows, and encourage them to earn points by using websites Twitter or Facebook to create online buzz for programs such as 'The Office' and 'Chuck'. [...] The Nielsen Company reports Americans spend about 3.5 hours each month simultaneously gazing at their TV sets and the Internet, which is a 35 percent increase from a year before (Dobuzinkis, 2010).

This increasing phenomenon is made of two related processes: a) the progressive integration between TV and social media in a new combined way of consumption; b) the way in which consumption is being brought to fruition and it becomes a promotional activity of people using the network to show up their passion and amplify the TV effects. The experience generated by this new integration can be handled by the TV brands or by other subjects only if the audience has the perception of an authentic relationship with the brand. In this way TV is changing its identity, turning it into a brand that works as a provider of experiences.

Many researchers today are focusing on the convergence between TV, blogs and Web 2.0 as the main horizon of innovation in this field: new frames, new contents, new roles of production and consumption. This is why we cannot talk anymore of simple single technologies but we must analyse the construction of a new "communicative environment" (Fortunati, 1998). In this environment Web 2.0 is not just a new tool or a new medium, it is more the general Internet which connects different media, from the strategic dimension (governed by broadcasters and corporations) to the tactic one (governed by users). So the older TV device can be re-generated in the new Web context. This could happen if TV is capable of upgrading itself to deliver the three main dimensions of experience, emotion and relation. TV can not be just an enclosed device which transfers contents from one to many. We are already involved in a process of co-design the world of TV in which the network dynamic will fulfil the gap between TV and Web 2.0. At the same time the logic of blogging and posting will probably generate a remediation between different technologies, or better what Manovich (2008) consider as a metamedium which is expanding the concept of TV fruition. As the social network is an extension of our identity, also the micro web personal TV is a narrowcast medium that extends the identity of the "producer".

The convergence between branding, self-mass communication and Web 2.0 opens a new frontier in which TV changes its original destination and become a way to augment the self and implement a sophisticated strategy of social positioning. This modality of an existential positioning comes from the blurring of the boundaries between public and private space made concrete by the transparency of the new media. We are probably in the age of "prosuming" (Toffler, 1979) but this definition doesn't clarify the way in which some productive and consumptive elements are recombined in their respective spheres. Better than this, it could be helpful to look also at Flichy's idea of "professional amateur" (2010) as the paradoxical effect of an even more elitist society which only tries to appear more democratic.

In other words the expression "Personal Web TV" is ambiguous, because it defines different meanings and objects. It can refer to a consumption protocol which offers to the audience a strongly customized table of contents, as in the case of Apple Web TV. On the other hand this definition involves the production contest or better brings the logic of customisation from the side of consumption to the side of production. In this case the definition indicates the way in which new platforms give the users the possibility of managing a real, owned TV channel.

In the case of consumption, marketing has developed several solutions to customize the TV offers of contents on the single profile of the user. A German research on cross-modal analysis of the TV news broadcast adopts a range of new techniques to study the production of a national broadcast such as face detection with a k-means algorithm, optical character recognition, speech recognition. The same type of analysis can be useful to define the customer consumption, in fact

> different practical scenarios for TV channel independent personal TV solutions were presented based on cross modal analysis (Dunker, Gruhne and Sturtz, 2008, p. 4).

If we shift over the cross-modal statistical analysis to a deeper phenomenological one we can find the qualitative specificity of this new technology compared to the old TV device. This is what a recent study on interactive design (Heeter, 2000) has developed on the comparison between TV and personal TV focusing the different affordances that characterize those different media. In this going "social" we can find a new way of naturalizing the bringing to fruition a new media that is so different from the modality of the old media (Heeter, 2000; Reeves and Nass, 1996).

Video on demand is a vastly different form of interactive television than chatting online in a corner of the screen with the star of a show while the show is on. Most of the interactive services involve more viewer activity of some kind, although the Personal TV Services may actually result in less viewer activity because the intelligence in the box does the programme selection for the user. Marketing each of these services as interactive television helps inform consumers that the service will be different from the traditional television viewing experience they are accustomed to. However from a research and design standpoint, a more clearer, better differentiated conceptualisation of interactivity is needed. Wrestling with this challenge eventually lead me to the transcendental phenomenology for fundamental definitions of human experience as a foundation for a theory of interactivity and our interface to the world (Heeter, 2000).

There is a complementary relation between the consumer experience and the production procedures. When the TV brand puts a part of its activities in the hand of the audience – as in the case of Twitter fans trying to rescue the "Ronna and Beverly" TV pilot – it cuts the distance between producers and consumers. This approach is the precondition of a straight integration between them. At the same time, the audience becomes "productive" taking part in the TV brand choices and the TV brand becomes a "friend" like others.

Many innovations are moving in the sense of an increasing power, which it puts in the hand of the users. There are several experiments or spin-off coming from the academic research, trying to implement a new model of "self-television". In the grassroots dimension (Jenkins, 2006) we can reflect on several amateur cases which usually overlap with the micro or neighbourhood TVs. There are also experimental projects to create an infrastructural service for people who want to launch their own personal Web TV such as TV4U, a Japanese researchers spin off. Hamaguchi, Doke, Hayashi and Yagi (2005) start from the evidence of the high level of diffusion reached by blogs and UCG (User Content Generated) via social networks. In the following quote we can see how their work hypothesis a personal Web TV as a "molecular" and spread form of production, could be a future scenario.

Although low-cost video cameras and desktop editing tools are popular consumer items, we are scarcely able to see personal TV programmes made by non-professionals. It seems reasonable to conclude that making TV programmes is still quite difficult for ordinary people just using consumer production tools. Looking at the explosive growth in popularity and the spread of blogs over the last few years, indeed, we would expect to see the same kind of remarkable growth in personal TV program and content production if user-friendly tools were available to easily create and distribute TV program content (Hamaguchi, Doke, Hayashi and Yagi, 2005, p. 60).

TV4U tries to customize the scripture of TV programs operating on the main three levels of program production: a script editor, a previewer and an uploader. In few words it is the implementation of:

> user-friendly system enabling users to produce and distribute their own personalized TV programs over the Internet that can be easily accessed and viewed by others (Hamaguchi, Doke, Hayashi and Yagi, 2005, p. 60).

A similar technology opens a new frontier for the development of new UGC television. However at the moment it is probably more useful to study the present state of these phenomena.

Web TV as a Brand Name Tactic Extension: Simona Ventura's Personal Web TV

As usually happens with digital innovation and its cultural implications, the grassroots movements – based on a low budget investment – are turned into a new powerful tool by corporations, institutions or famous personalities. This movement from bottom to the top, from local to global and from consumption to production, is the way in which original needs coming from the people are subtracted and counterfeited to be used from the top to the bottom as an expression of power.

In the first case, users use new media to communicate on a larger scale and to turn themselves in something more complex than just a user: a prosumer, a content generator user, a prof-amateur. In the second case, TV channels and celebrities are seeking new market niches and more diversified audiences and creating new brand extensions of their core activities. In both cases the communicative process is focused on the protagonist's real life but in the first case it is generally managed directly by the person who elaborates the project, such as a more complex videoblog. In the second case the famous person becomes the protagonist and the subject of a representation managed by a professional teams (e.g. producers, PR experts, press office etc.). In this final part of the paper I will focus on the second cases, trying to compare three different national initiatives in: USA, Germany and Italy.

In the last few years there was a wide expansion of the micro TV's phenomena with several international cases representing this new way of production which is, in a sense, the overturning of what Jenkins defines as a "convergence culture" (2006). Here, in fact, the convergence between languages and media is not properly developing on the side of consumption but rather it is the way in which the consumption culture is embedded in the productive sphere, as the easiest way to streamline the production.

A famous example of this process is the Oprah Channel on YouTube, which is the online projection of her TV Show even if less structured than OWN (Oprah Winfrey Network), a meaningful acronym which underlines how this channel is dedicated to the empowerment of her public figure. The Oprah's channel on g is the online extension of the powerful identity that the most famous American TV presenter gained with her show. It is no coincidence that Eva Illouz (2007) also considers the famous Oprah Winfrey's Show as the natural landing of a long distance journey started in the American industries during the 30's. Out of this history was born the dynamic of the "emotional ontology" which quickly turned into "emotional capitalism" during the twentieth century. When this new form of capitalism moves from the real enterprises to the fictional world of TV, it unveils a new way to communicate with the audience. In other words, after big cultural or counter-cultural movements such as the self-help or the feminism, the emotional capital becomes a fundamental resource of the new television system. It opens a new kind of social dimension that Bauman has recently defined as "confessional society" (Tester, 2001). Confession is in a way the trait d'union between generalist and new media, or just the way in which TV prepares its decline and reborn through the diffusion of the Web based forms of new TV.

Today, in the Web 2.0 era, the same confessional issues shown in the past are reinforced and more referenced to the presenter figure. The presenter becomes the centre of a crossmedia strategy which integrates several activities and makes them "social". The official Oprah website is not just a promotional tool, it is more a platform which extends the contents of her Show and integrates its activities with the Oprah official YouTube channel, the OWN TV and also the Magazine and the radio. Each programme of her TV (e.g. Why Not?, Finding Sarah, or Unfaithful) has a space for discussion with clear rules to select the participants. The website also has a casting space launching the audience into the "OWN experience". Every box shows a category and the criteria of selection. One of them, for example, is dedicated to the research of mothers with "extreme lifestyle choice" the website asks sons and daughters: "Is your mom a workaholic, shop-a-holic, or does she have compulsive spending problems? Does your mom have an obsession with plastic surgery, or is she addicted to exercise? Or, is your mom a 'cougar'?" The logic of the self-discovery is diffused on a micro-physical level, involving all the generations.

The sum of YouTube web channel, the official website and OWN generates a field of interaction with users and compensates for the perception that Oprah's brand may be seen not only as a mainstream phenomenon. This integrated multimedia empire of communication is quite different from the small Boris Becker's Web TV, which is much more focused on the strongly performing image of the most famous eighties tennis champion. Even if the sport is the main content of his TV it can be said that there are other issues used to empower the public image of the star. Family, everyday life scenes and fans are equally relevant details that help to make the eccentric image of the celebrity more human and mature. The real peculiarity of this TV is the fundamental role of Becker's wife, Lilly, in the construction and representation of his public image. The point in the menu where her name is listed is one of the six main categories of navigation: home, sport, lifestyle, Lilly, charity, entertainment. The centrality of her position is as if the entire operation is built around her. Perhaps, as in the past, a few magnificent monuments were built to celebrate the love of a king for his wife, also in this case the celebration of Lilly sounds like a contemporary, virtual and living monument. For many reasons the Boris Becker's Web TV is not comparable to the Oprah's YouTube Channel. In fact Becker is not the same top star of the past so this new tool of communication is used to keep alive his image and iconography, extending his brand identity from the sports to a more general skill.

The project of the personal Web TVs seems to be shaped on the same strategic needs of the contemporary commercial brands. The notiont that advertising (and broadcasting as well) suffer from a "lack of credibility" today, is the point of reflection of A. and L. Ries (2002). These authors, in fact, try to demonstrate that public relations (PR) are much more useful than advertising if you want to build a brand, even if, in the reality, we must say that there is no such contradiction between them. In fact there is more a complementary relationship so that "advertising should continue to reinforce PR ideas and concepts" (p. 197). This is way the personal Web TVs are usually positioned between mainstream TV and more tactical media such as the social networks. Because they are much closer to the niches, rather than the traditional TV brands that are more related to the mass market and to the necessity of the advertising approach. In other words their business model has nothing to do with renting spaces, rather it is more focused on the creation or the empowerment of the celebrity brand name.

The main Italian example of this personal Web TV diffusion is Simona Ventura's recent experiment which is a dialogue between different media. This project is smaller than Oprah's multimedia empire and bigger than Becker's Web TV but it is very representative of how the Web TV could be used as a project of image regeneration and self re-launch in celebrity marketing. Ventura is definitely the most famous Italian TV presenter of some very popular programmes of the Italian public television but recently her role in national show business started to decline. Not only for this reason is her case is similar to Boris Becker, in fact she also established her career via sport as a journalist and then married a not particularly talented, some would say, but famous football player. From that moment she started to use sports as a powerful trait of her identity in Italian show business. Her Web TV is not very structured on the productive and technological side. It is more a tool which creates new forms of interaction with the audience, using the connection with her own public relations network. In this way it works similarly to a Facebook page but also offers several types of content, most of which are related to her mainstream television.

Recently, her Web TV attracted the public attention through three initiatives. The first regarded a famous reality show presented by her: "L'isola dei famosi" (the Italian edition of Charlie Parson's Celebrity Survivor!). Aiming to break down the barrier between the studio and the "real scene", she announced her intention to land with a parachute on the island in Honduras. This event was totally covered by her Web TV which capitalized the attention of the Italian press, the pick of the audience and the high notoriety of this mainstream TV programme. The second initiative was the interview with one of her close friends, Lele Mora, who was under investigation for his involvement with the recent sex scandal of The Prime Minister Silvio Berlusconi. Mora tried to use this tiny and cosy media space to advance his public defence, helped by Ventura's high social respectability. She interviewed Mora as a "victim" in an intimate and friendly confessional style. This was before his legal case was damaged by the publication of his wire-tapped conversations with the other protagonists of the scandal (Donadio, 2011). The third big event was the interview of the famous crew of Italian-American girls and boys, protagonists of the American MTV reality "Jersey Shore". Ventura had a chance to interview them during a trip to Florence, where this sort of "diasporic community" (Appadurai, 1996) was trying to reconstruct its ties with its country of origin. It exposed instead an evident contrast between the cultural heritage of the city and their lavish and trashy style (Pisa and Adam Otis, 2011). Both cases are a significant example of how the Web TV, focused on the TV celebrity, feeds many forms of interdependence. In the first case, blurring barriers between the studio and the island, it puts the presenter in the centre of the "real" action. In the second case, there is a process of reciprocal support between different Italian celebrities. In the third case, the personal Web TV gives to the presenter the freedom to interact personally and confidentially with celebrities coming from other countries and working for other competitor broadcasters (since "Jersey Shore" is aired on MTV Italy).

Discussion and Final Remarks

According again to Abruzzese's and Borrelli's classification (2001), these new forms of entertainment and their aesthetics, look like an extreme manifestation of the post-television era. There is no ideology governing the selection of contents, neither is there a purely commercial need. Here there is just the hypertrophic ego of the presenter, able to act and move in total freedom, so it becomes in a certain sense the medium. The consumer experience is completely monopolized by the identity of the presenter. In a paradoxical way, the TV remediation of television in a narrowcast dimension, recreates a new form of hegemony which is closer to a project of total manipulation of the old television models. Personal Web TV represents the extension of the celebrity brand name and follows what in the eighties was the ideal of "total look": a complete occupation of the consumers identity, imposed by the fashion brands.

The recent theoretical concepts such as "self-mass communication" (Castells, 2009), "self-packaging" (Siegel, 2008) or "self-branding" (Barile, 2009), underline a new liberation of digital media users but also a new form of hegemony. The age of the glamorous TV is probably ended. Thus the contemporary V.I.Ps. are using new tools to show up and regenerate the fundamental value of the new show business: authenticity. Web 2.0 recreates a space of direct interaction between the celebrity's body and the public common experiences, fulfilling the gap between the show and the reality of life. Authenticity is the new lifeblood which feeds both show business and consuming processes. In other words we are moving towards a stage where TV brands, celebrities, logos and designers are less important than "memorable experiences" and reality counts more than a simple fiction.

> All human enterprise is ontologically fake [...] and yet, output from that enterprise can be phenolmenologically real – that is perceived as authentic by individuals that buy it (Gilmore and Pine, 2007).

The evolution from the old broadcasting strategies to the new media tactic means a new way in which celebrities can be represented as common "mortals" and share their own experiences with the public. These new trends in TV entertainment are based on the "soft substance" of human experiences, relations and emotions and are not just related to the diffusion of a global fan culture. It is more about how fans can be involved in a general logic of self-branding and prepare themselves to be the producers and the actors of their everyday life, through a new distilled form of "storytelling" (Salmon, 2007). In other words we are passing through the traditional TV regime, based on the traditional show system rules, approaching a new regime made up of digital innovation and based on the public circulation of narrated authentic emotions.

References

Abruzzese A., Borrelli D. 2001. L'industria culturale. Tracce e immagini di un privilegio. Roma: Carocci.
Agamben G. 2006. Il dispositivo. Roma: Nottetempo.
Appadurai A. 1996. Modernity at Large. Cultural Dimension of Globalization. Minneapolis: Minnesota University Press.
Barile N. 2009. Brand New World. Il consumo delle marche come forma di rappresentazione del mondo. Milano: Lupetti.
Baudrillard J. 1994. The Illusion of the End. Stanford: Stanford University Press.
Bolter J. D, Grusin R. 1999. Remediation: Understanding New Media. Cambridge: MIT Press.
Bourdieu P. 1984. Distinction: A Social Critique of the Judgement of Taste. London: Routledge.
Brooks D. 2000. Bobos in Paradise: The New Upper Class and How They Got There. New York: Simon and Schuster.
Castells M. 2009. Communication Power. Oxford, New York: Oxford University Press.
de Certeau M. 1984. The Practice of Everyday Life, translated by Steven Rendall. Berkeley: University of California Press.
Donadio R. 2011. Wiretaps of Berlusconi's Teenage Friend Emerge. The New York Times, 18 January.
Dobuzinskis A. 2010. NBC gets Social in Web-based Viewership Push, Los Angeles: Reutgers, 13 May.
Dunker P., Gruhne M., Sturtz S. 2008. Personal Television: A Crossmodal Analysis Approach, IEEE International Symposium on Consumer Electronics, Conference paper: Volamoura Portugal.
Eco U. 1985. La trasparenza perduta. In: Sette anni di desiderio. Milano: Bompiani, pp. 163-180.
Eliacheff C., Soulez Larivière D. 2007. Le temps des victims. Paris: Albin Michel.
Flichy P. 2010. Le sacre de l'amateur. Paris: Seuil.

Fortunati L. (ed.). 1998. Telecomunicando in Europa. Milano: Franco Angeli.
Foucault M. 1978. The History of Sexuality. Vol. I: An Introduction, translated by Robert Hurley, New York: Pantheon.
Gilder G. 2000. Telecosm: How Infinite Bandwidth Will Revolutionize Our World. New York: The Free Press.
Gilmore, J. H., Pine B. J. 2007. Authenticity. What Consumers Really Want. Boston: Harvard Business School Press.
Hamaguchi N., Doke M., Hayashi M., Yagi N. 2005. Internet-Based Personal TV Station Enabling Individuals to Produce, Distribute, and View TV Programs (full paper), The IADIS International Conference WWW/Internet 2005, Proceedings Vol. 1, pp. 52-60.
Heeter C. 2000. Interactivity in the Context of Designed Experiences. Journal of Interactive Advertising, 1 (1), Fall.
Illouz E. 2007. Cold Intimacies: The Making of Emotional Capitalism. Oxford: Polity Press.
Jenkins H. 2006. Convergence Culture: Where Old and New Media Collide. New York: New York University Press.
Kay A., Goldberg A. 1977. Personal Dynamic Media. Computer, 10 (3), pp. 31-41.
Lyon D. 1994. The Electronic Eye: The Rise of Surveillance Society, Minneapolis: University of Minnesota Press.
Manovich L. 2008. Software Takes Command. Released under CC license. URL: http://softwarestudies.com/softbook/manovich_softbook_11_20_2008.pdf (accessed 10 July 2011).
Mason M. J. 2008. The Pirate's Dilemma: How Youth Culture is Reinventing Capitalism. New York: Free Press.
Mattiacci A., Militi A. 2011. TV Brand. La rivoluzione del marketing televisivo. Bologna: Fausto Lupetti Editore.
McLuhan M. 1964. Understanding Media: The Extensions of Man. New York: McGraw-Hill.
Mayer-Schönberger V. 2009. Delete: The Virtue of Forgetting in the Digital Age. New Jersey: Princeton University Press.
Mirzoeff N. 2000. An Introduction to Visual Culture. London et al.: Routledge.
Pisa N., Adam Otis G. 2011. Florence rips 'Jersey Shore' supercafoni (Translation: Big Idiots). New York Post, 28 May.
Reeves B., Nass C. 1996. The Media Equation: How People Treat Computers, Television, and New Media Like Real People and Places. Cambridge: Cambridge University Press.
Ries A., Ries L. 2002. The Fall of Advertising & The Rise of PR. New York: Harper Business.
Salmon C. 2007. Storytelling. La machine à fabriquer des histoires et à formater les esprits. Paris: La Découverte.

Schmitt B. H. 1999. Experiential Marketing. How to Get Customers to Sense, Feel, Think, Act, Relate to your company and brands. New York: Free Press.
Sennett R. 2008. The Craftsman. London: Allen Lane.
Siegel L. 2008. Against the Machine. How the Web is Reshaping Culture and Commerce – And Why it Matters. New York: Spiegel & Grau.
Tester K. 2001. Conversations with Zygmunt Bauman. Cambridge: Polity Press.
Toffler A. 1979. The Third Wave. New York: Bantam Books.
Uricchio W. 2010. TV as Time Machine: Television's Changing Heterochronic Regimes and the Production of History. In: J. Gripsrud (ed.). Relocating Television: Television in the Digital Context. London: Routledge.
Vincent J., Fortunati L. (eds.). 2009. Electronic Emotion. The Mediation of Emotion via Information and Communication Technologies. Berlin: Peter Lang.

Emiliano Treré & Manuela Farinosi

(H)earthquake TV: 'People Rebuilding Life after the Emergency'

Introduction

According to several studies, traditional Italian television framed the post-earthquake situation as a "miracle in L'Aquila" and a spectacular of the pain of those involved in the earthquake to impress the TV audience without respect for the feelings of the victims. In contrast to this situation, it seems that, on the Web, the victims of the tragedy could speak with their own voices without any mediation.

The aim of this chapter is to explore how the dimension of pain was represented on a cross-media platform – FromZero TV – that documented the lives of the people living in the tent camps after the earthquake tragedy. Using qualitative methodologies (video analysis and interviews), we found that this Web TV offered a considerate and balanced representation of pain in relation to the victims of the catastrophe. It also emerged that this platform was able to represent grief in a more respectful way adopting the same words and terms of the affected population, without having to stick to the rules of the traditional television agenda.

From Zero TV thus represents an interesting "experiment" in how new forms of television on the Internet can offer alternative representations of events and give a voice to ordinary people without having to appeal to the exhibitionism of feelings or to stick to the rules of the traditional media agenda.

Communicating Disasters

After the L'Aquila earthquake in Italy in 2009, the need to spread and receive information by citizens, communities and public institutions increased dramatically. The quake thus constitutes a benchmark for the analysis of the multiple voices that comprise the Italian media scenario. Research has shown that in tragic situations, such as hurricanes, floods and earthquakes, traditional media, and increasingly new media as well, play a critical role and act as crucial management tools, able to transmit official information about the emergency in different ways. Their organizational and institutional characteristics make them particularly well suited to information gathering and communication during natural disaster events (Miller and Goidel, 2009). Furthermore, catastrophes tend to produce a strong impact on the whole media system and on all the other communication channels, forcing them to react faster and with more emotional involvement than in a "normal" situation (Burkhart, 1991; Masel-Walters et al., 1993; Morcellini, 2006; Tota, 2003). This branch of literature specifically focused on disasters underlines that the media agenda tends to change in relation to the three phases that characterize a critical event: impact phase, emergency phase, reconstruction phase. These three phases are usually mirrored in the media coverage of the disaster. The media play a critical role since they collect information as the disaster unfolds and provide a channel for public officials to communicate with the public about the management of public organization. Furthermore, they play a pivotal role in framing the disaster as a news story, giving the story broader cultural resonance and political and social meaning. As Garnett and Kouzmin (2007) have pointed out, it is difficult to imagine governments effectively responding to a natural disaster in the absence of news organizations reporting breaking events.

The media are favourably situated to gather and transmit information about natural disasters, which can help citizens and policy-makers understand the scope, causes, and consequences of a catastrophe. But the media are, at the same time, subject to institutional biases that may lead to distorted presentations of reality and perpetuate misinformation, stereotypes, and misunderstanding (Miller and Goidel, 2009). Chomsky (1998, p. 42) asserts that expecting the media to report disasters in a more honest and humane way on their own initiative – rather than reflecting the interests of the powerful – would be like "expecting General Motors to give away its profits to poor people in the slums".

It has been highlighted that public officials engage in communication strategies designed to promote favourable narratives about the causes, consequences and lessons to be drawn from the disaster (Boin, 'T Hart and McConnell, 2009).

In a review of existing disaster research, Quarantelli (1991) pointed out that:

> Mass media reports, especially in television, tend to present content that perpetuates certain disaster myths. [...] For example, although references to panic and looting constitute only a small proportion of the total television content, their presentation is very dramatic and consistent with the mythologies (p. 39).

Furthermore, from a review of the literature on disasters and communication, it emerged that while there is quite a long tradition of studies and analyses regarding traditional media, such as television and radio (Burkhart, 1991; Dayan and Katz, 1992; Lombardi, 2005; Masel-Walters et al., 1993; Morcellini, 2006; Pasquarè and Pozzetti, 2007; Perez-Lugo, 2001, 2004; Ploughman, 1997; Quarantelli, 1991; Scurati, 2003; Tota, 2003; Turner, Nigg and Paz, 1986), there is also a growing amount of research on how Internet technologies are used before, during and after catastrophes (Garnett and Kouzmin, 2007; Macias, Hilyard and Freimuth, 2009; Meier and Munro, 2010; Muralidharan et al., 2011; Skinner, 2004; Yates and Paquette, 2010).

Another branch of literature it is useful to look at is the media representation of disasters. Media events are defined by Dayan and Katz (1992) as public ceremonies, deemed historic and broadcast live on television. The authors refer to the exploitation by the political systems of televised live, pre-planned events, such as the funeral of President Kennedy or the Olympic Games, to celebrate and at the same time reproduce the social system. Recently, answering the criticism of having avoided in their analysis news events that shock the world (Scannell, 1996; Katz and Liebes, 2007) have revised the original Dayan and Katz's analysis and have distinguished between "media events" and "disruptive events", such as disasters, terrors and wars. The two authors assert that ceremonial events are in decline, while the live broadcasting of traumatic events and the obsessive coverage given to them by mainstream media has dramatically risen. Liebes (1998) calls these broadcasts "disaster marathons", referring to the hours and days that the media spend recycling gory portraits from the scene, highlighting the heroics of rescue workers, and carrying out interviews with experts and politicians, speculating on the reasons for the disaster.

Katz and Liebes (2007) link the decline in salience of the live broadcasting of ceremonial events and the consequent rise in the live coverage of traumatic events on one side to an increasing cynicism towards establishments and media, and on the other to new and complex media ecologies that have

> scattered the audience and undermined the shared experience of broadcasting, have taken the novelty out of live broadcasting and have socialized us to 'action' rather than ceremony, to a norm of interruption rather than schedule (p. 159).

In recent years, television has undergone strong changes, especially related to the process of digitalisation (Treré and Sapio, 2008; Treré and Bazzarin, 2011). While television has had, for more than five decades, a precise and relatively stable status and the "television experience" could be defined in a rather clear way (Scaglioni and Sfardini, 2008), strong processes of innovation, diversification and hybridisation are reshaping the current framework of this medium and the new opportunities supplied by the Internet (Colombo, 2004; Fortunati, 2008) are changing and shaping the traditional television scenario, often in unpredictable ways. The actual media scenario is composed by multiple communication technologies and platforms, with several crossovers and influences existing among the Internet and "traditional media" such as television (Fortunati, 2005; Treré, 2008). The phenomenon of media convergence is not only a technological one, but mainly a cultural one. Henry Jenkins (2006) has coined the expression "convergence culture" to refer to the changes which new media bring about in our society at a broader, cultural level. We can thus speak of convergence television (Grasso and Scaglioni, 2010) where technologies, broadcasters, users, programmes and practices mutate and redefine each other because of the interrelations between the traditional media and the emergent new media ecology. Television experience is broadening and multiplying, both in quantitative terms (becoming more available) and in qualitative terms (through a process of personalization). We may thus frame the actual television scenario inside a continuum: on one side, the traditional broadcasting, in the middle different enhanced forms of television, moving towards the other side, where television hybridises with the Internet (Scaglioni and Sfardini, 2008).

In this chapter, we first present a brief overview of the L'Aquila earthquake, analysing some of the most relevant episodes that characterized the media coverage of the earthquake, and then introduce our case study. Secondly, we introduce our research methodology, then we report and discuss the findings of our analysis of the videos and from the interviews. Finally, we draw some conclusions from our research, critically reflect on our findings, and propose a few important topics that deserve further research in the future.

L'Aquila Earthquake and the Media Representation of Pain

On 6 April 2009 at 3.32 a.m., a 6.3 Mw magnitude earthquake struck in L'Aquila, a small Italian city (around 75,000 inhabitants), capital of the mountainous Abruzzo region, located approximately 85 km north-east of Rome. The first earth tremor was followed by two large and serious aftershocks on 7 April (Mw=5.6) and 9 April (Mw=5.4). This catastrophe represents Italy's worst earthquake in the last 30 years and the deadliest since the 1980 Irpinia earthquake. Earthquakes have marked the history of this place; in February 1703, L'Aquila had already been almost destroyed by the biggest earthquake experienced by the Italian peninsula in the previous century. That quake killed around 5,000 people and destroyed much of its medieval historic centre, which was then rebuilt in the Baroque style. So the April 2009 quake was not the first to strike the central Italian city: seismic activity is relatively common in the Italian peninsula, but luckily intensity like that of L'Aquila's earthquake is rare. The media response to this event was immediate and expensive. Douglas Kellner (2010, p. 76) refers to

> media constructs that are out of the ordinary and outside habitual daily routine which become special media spectacles. They involve an aesthetic dimension and are often dramatic.

For him

> media spectacle refers to technologically mediated events, in which media forms such as broadcasting, print media, or the Internet process events in a spectacular form (Kellner, 2010, p. 76).

Kellner grounds his theory in Guy Debord's notion of the society of the spectacle and in Dayan and Katz's notion of "media event". He distinguishes between spectacles of terror (such as the 9/11 attacks on the Twin Towers) and spectacles of catastrophe, which refer to natural disasters like the Asian Tsunami or Hurricane Katrina.

The academic literature on the media coverage of the L'Aquila earthquake (Dominici, 2010; Farinosi and Treré, 2010a, 2010b; Imperiale, 2010; Padovani, 2010) agrees on stressing the construction of a media spectacle framework by the Berlusconi government and on highlighting the differences between the many inflated promises of the Italian government about the reconstruction process and the sad reality of a city left alone when the media lights had disappeared. Moreover, there is ample journalistic literature on the topic (among others: Bonaccorsi et al., 2010; Ciccozzi, 2010b; Puliafito, 2010) where the L'Aquila earthquake is viewed as a spectacle of catastrophe, and the immediate post-earthquake emergency is seen as being framed by Italian mainstream media within the metaphor of the "miracle in L'Aquila". This strong metaphor is based on the use of several mechanisms and techniques, derived from the advertising and marketing spheres, to create image events in order to convey the false idea of a miracle related to a supposedly fast and efficient post-quake reconstruction.

Berlusconi appeared several times on national and international television screens: speaking from the "red zone" (a term designating unsafe areas of the historical city centre), he reassured the crowd that the city was undergoing a fast reconstruction process. Through a myriad of image events and inaugurations of new houses for the population culminating in the G8 Summit (Padovani, 2010), the Italian prime minister transmitted to Italian and international audiences the image of a city that was undergoing a strong and almost "miraculous" process of (re)-construction. But for many people of L'Aquila it was nothing more than a media spectacle, another example of mystification from the media tycoons that was in contrast with the sad reality of their everyday existences, where rubble dominated the centre of the city and lots of people were still living inside the tent camps (Farinosi and Treré, 2010a, 2010b; Imperiale, 2010).

Journalists and commentators have viewed the media coverage of the L'Aquila earthquake as an example of biased representation of the reality of the tragic event (Bonaccorsi et al., 2010; Puliafito, 2010) which promoted a favourable pro-government narration. Commentators have underlined that mainstream media, and in particular Italian television, have focused mainly on the emotional side of the tragedy (Dominici, 2010), using "sensationalist" and "triumphalist" ways of narrating the stories related to the dramatic event (Barile, 2009; Ciccozzi, 2010a). In particular, traditional television portrayed the pain of the victims of the earthquake several times, emphasizing the spectacular side of the dramatic event, without respecting the trauma and the suffering of the people involved. We refer here to the numerous images and photographs appearing in the days immediately after the earthquake that, in many cases, were not aimed at documenting the disaster, but lingered on scabrous details. These images did not add anything in terms of information, but only responded to an almost obsessive need to show the audiences the catastrophe according to the traditional canons of disaster movies (Dominici, 2010). All these elements have constituted the basic components of the media representation of the quake: the excessive insistence of some journalists to ask inappropriate and inadequate questions, the lack of sensitivity in taking the cameras everywhere and in every circumstance, recording the desperation of survivors and of the relatives of the victims, the despair of those who had had their houses completely destroyed by the quake, the injured taken away by ambulances, the dead bodies pulled from the rubble. This media spectacle was built without any consideration and respect for the involuntary protagonists of the tragedy and with little awareness of the nature of the situations.

The images on television often focused on that which attracts the audience attention, on that which touches, moves, distresses, worries. In many cases, a certain way of "making news" has, in the name of the right to inform, produced an information genre than is a "hybrid", between fiction and "reality show". Images, comments, tears, cries of despair, moments of suffering from the "private" sphere have thus become "public", going beyond the sacred right to inform and be informed and feeding a morbid curiosity to watch "with your own eyes" the suffering and pain of others.

The creation of a spectacular of the tragedy went hand in hand with the exploitation of the pain of the victims. Two episodes are particularly relevant in this respect. The first one is related to the most important Italian television news, TG1 (Farinosi and Treré, 2010b; Puliafito, 2010).

On the TG1 evening edition, broadcast on RAI channel 1 the day after the quake, the journalist started the programme by reporting the incredible and unprecedented audience share that TG1 had received due to its coverage of the earthquake. TG1 celebrated a private "audience victory" without respect for the victims of the tragedy.

The second example refers to the night of 8 April 2009. A journalist from the information television show "Matrix", broadcast on TG5, Italy's most viewed private television channel, started to knock on the doors of the cars where the victims of the earthquake were sleeping because they had lost their homes. The journalist entered with her microphone and, with the light of the camera inside the cars during the night, asked the people why they were staying in their vehicles. People were shocked and some were almost blinded by the light of the camera, while others were shouting that they had already talked to journalists and they wanted to be left in peace with their families. These are only two examples of how mainstream television made a spectacular from the pain of those involved in the earthquake, to impress the audience without respect for the feelings of the victims.

These mechanisms have paradoxically created a marked difference between the performance of the tragedy and the tragedy itself, between the real-life experience of those involved in the earthquake and the media spectacle of the catastrophe (Kellner, 2010).

Furthermore, Italian media, and television in particular, focused their coverage on the earthquake, especially in the first weeks afterwards, but they have not documented and reflected on the slow and difficult process of reconstruction of the ordinary citizens' daily lives. While Italian television and mainstream media have largely neglected the stories, contradictions and daily difficulties of the people of the tent camps and often exploited their emotions, the multiple voices of the citizens of L'Aquila have found their space on the Internet. Two decades ago, research had already pointed out that, after disaster situations, electronic media can bypass the gatekeeping process (Quarantelli, 1991).

Using websites, blogs, social network sites and video/photo sharing platforms, many ordinary citizens of L'Aquila were able to post articles, videos and pictures, to cast light on the post-quake situation in the fullness of its aspects and nuances, provide their own reflections on the tragedy, narrate their everyday lives (Farinosi and Treré, 2010a; Micalizzi, 2010) and use Internet platforms in order to create an alternative public sphere (Farinosi and Treré, 2010b). These people have used the Web as a powerful infrastructure to bypass traditional gatekeepers (Bennett, 2003; Rucht, 2004) and directly communicate their critical views on their post-quake situation. Moreover, activists, collectives and civic movements (such as, for instance, the movement of the "People of the Wheelbarrows" and the "3e32" collective) have employed multiple Internet technologies including Web 2.0 platforms like Facebook, YouTube, Twitter, Flickr, WebTVs and blogs to organize and report citizens' protests and to provide their perspectives on the post-quake events in L'Aquila (Padovani, 2010; Farinosi and Treré, 2010b).

Case Study: FromZero TV, a New Global Web Platform to Narrate Life after the Earthquake

Acknowledging that Italian traditional television often represented the pain of the victims of the earthquake in spectacular ways, in this contribution we shift our attention to the Internet and in particular to a cross-media platform, FromZero TV[1], created some months after the quake to narrate the multiple stories of the earthquake survivors, focusing on their efforts to rebuild their lives after the catastrophe.

The FromZero TV project has been conceived by two independent Italian movie producers: Stefano Strocchi's Move Productions and Pulse Media. The project has involved three documentary directors, each coming from a different part of Italy to live and work in the tent camps. In addition, four editors lived in the

1 http://FromZero.tv/eng.html

camps and edited the episodes for the Web (webisodes) while the directors were shooting. The crew was hosted by the Italian Red Cross in the "Centi Colella" tent camp. FromZero TV was realized with the support of Al Jazeera English, the Italian Red Cross, the Intesa San Paolo Bank, the Torino Piemonte Film Commission and sponsored by the Province of L'Aquila, the Abruzzo Region and the Abruzzo Film Commission. The realization of this platform was inspired by the "Gaza-Sderot" project which was mainly based on the idea of combining the Internet with the making of documentaries.

The team started shooting on 1 September 2009 and lived in the camp for over two months. The aim of the FromZero TV platform was to follow the everyday lives of the people in the tent camps "as they were slowly building a new beginning" (quote from the FromZero TV website), therefore shedding light on the happenings in a disaster area when there is no more media attention and the slow process of "a return to normality" has begun. The purpose of FromZero TV was to video-document and narrate the multiple stories of the earthquake survivors, focusing on their efforts to rebuild their lives after the catastrophe. The guiding idea of the project was to transmit the feelings and the emotions that people experience when faced with a tragic event which forces them to redefine their existence. The creators of this platform have focused on aspects of the L'Aquila tragedy that Italian mainstream media have neglected, following some "characters" during their lives in the tent camps. According to Stefano Strocchi, FromZero TV's author and series producer:

> In Italy the news on TV was all about the political use of the response to the emergency, with a great debate about it, but no-one was paying attention to the stories of people, the volunteers, and everyone that, instead of debating, was there trying to overcome the trauma and to move on (Strocchi, 2009).

FromZero TV is configured as a form of cross-media television, because it was conceived from the beginning to be spread both on the Internet and on traditional television. The project was "Internet-first", because the main goal was to exploit the Web's immediacy and 24 hour a day availability, so television episodes were shot in the form of so-called webisodes, that is short videos (around 3 minutes), uploaded on the FromZero TV website and then spread on several blogs and multiple platforms. But the series was also conceived from the beginning to be broadcast on traditional media and two 22 minute films were then realized and broadcast on the Al Jazeera English channel. The first film, "FromZero TV: Ask the Dust" explores three of the L'Aquila stories and focuses on the tent camp, seen as a sort of "limbo". The second 22 minute movie, "FromZero TV: Next Moves", focuses on the survivors' attempts to return to normality, facing the task of rebuilding their city and their lives. The digitalization of the content has allowed the FromZero TV creators to remix and adapt the content for different platforms; currently, the FromZero TV producers are working on the release of a DVD of the series.

Aim and Methods

The aim of this research was to investigate how the pain of the victims of the L'Aquila earthquake was represented on the FromZero TV platform. Our research question was: has the FromZero TV cross-media platform been able to represent the pain of the victims of the earthquake in sober, balanced and respectful ways?

Our hypothesis was that the platform's documentary style with a strong focus on survivors' stories, together with the possibility of this web platform to bypass traditional gatekeepers have contributed to offer a coherent and more nuanced representation of the pain of those involved in the earthquake "in their own terms": closer to their everyday reality and less in the manner of a spectacular.

We decided to adopt qualitative methodologies because of the attention they assign to actors' perspectives, perceptions and understandings of the social world (Fontana and Frey, 2005; Patton, 2002; Vincent and Fortunati, 2009). Our first step was to conduct an exploratory analysis of the videos available on the FromZero TV platform. By using these videos as a unit of analysis, we aim to understand the types of content and the types of characters identifiable in this kind of television. We analysed the content of 41 videos: the average length of the sample videos was around 3–4 minutes. The videos were often filmed in high quality appealing to diverse audiences through the use of English subtitles. The plots are simple and focus on few topics. The videos were organized into two sections: episodes (N=29) and characters (N=12).

Our second step consisted of conducting 15 semi-structured interviews with 15 people – eight females and seven males aged between 21 to 47 years – who have lived the tent-camp experience in different campsites all around L'Aquila. The interviews began by showing to the interviewees the webisodes, uploaded on the FromZero TV platform. Two other two topics were also addressed: the pros and cons of this new form of television, and FromZero TV's ability to capture and represent the dimension of pain. Interviews were transcribed using the F4 software and thematically analysed (using the TAMS Analyser software) following Flick's (1998) method of thematic coding.

Findings

From the Analysis of the Videos

As already mentioned, FromZero Web TV comprises two interconnected sections: episodes and characters. The episodes amount to 29 and include 12 different characters. On the right of the website is found the characters section which contains a series of videos (N=12) focusing on the people of the tent camps who were chosen as "characters" from the FromZero TV crew because of their active roles in the community of the Centi Colella tent camp. These characters include people who were helping the survivors to regain normality: psychologists, Italian Red Cross volunteers, the chief cook, and the rest. Moreover, FromZero TV made videos of those involved in the earthquake and focused on their daily activities, that is the first day of school, visits to their collapsed apartments, the struggle to find new jobs, the harsh conditions in the camps, the attempts to rebuild a social structure, and so on.

FromZero TV's creators have highlighted that, even if the first phase of a catastrophe, like the earthquake in L'Aquila, can cause serious disruptive effects on the collective feelings of cohesion among citizens (Lavanco, 2003), in the second phase, such as in the L'Aquila tent camps, people develop a sense of community and start to slowly rebuild formal and informal systems of relationships together to regain their "normality".

As is stated in "the project presentation" section of FromZero TV's Internet site:

> Out of the private walls of the houses, forced into tents, the barriers of privacy and private life vanish, replaced by the need of sharing and mutual help. Through our series we'll paint a rough but truthful picture of human beings forced to live together, between everyday struggle, private selfishness, common interests, improvised solutions, rage, moments of celebration and humour as a community fights to regain normality (from the FromZero TV Internet site).

Shooting and uploading to the Internet the short episodes (N=29) showing the lives of those involved in the earthquake and their everyday activities, means also allowing

> everyone of us to be at the side of the survivors, to go beyond the news, into the everyday reality of people facing emergency and building from destruction (FromZero.tv).

The videos are short, simple and clear. They are shot in a documentary style without any voice-over or any interruption from the documentary makers and with just basic editing: the videos let the different characters speak about their experiences in their own words.

The focus of the videos is on the everyday activities of the people in the camps, and on the new relationships and new lives that they have tried to reconstruct after the earthquake. A wide variety of situations is portrayed in the episodes: a lawyer who has lost her office in the centre of L'Aquila and has to reinvent her life; a local artist who continues to work despite the tragedy; two friends whose friendship is strengthened by the earthquake; a group of children who are invited by educators to think about the city of their dreams during some recreational activities; an old woman who does not want to leave her house to go to live in a hotel on the coast; a woman who has decided to set up an improvised library inside the camp after her house was destroyed; a small child going to his first day of school in a temporary building managed by nuns and volunteers.

From the analysis of the content of the videos, it emerged that the dimension of pain in this difficult phase of reconstruction is clearly recognizable. Almost all the videos analysed contained some reference to the pain dimension, with some implicit allusions to sufferance. Most of the episodes do not treat the dimension of pain in an exposed way and do not make a spectacle out of it. Pain is present, but it is narrated without any particular connotation, in a sober and balanced manner. The people in the videos are not pushed by any external agents, they are free to express their feelings in front of the camera in their own terms and show their pain after the tragic event without having to stick to the rules of a traditional television agenda.

After the first days of seemingly unbearable pain that the citizens of L'Aquila had to confront, the people of the tent camps filmed by FromZero TV had somehow managed to deal with the pain of losing the houses, jobs, friends and the "normal" lives they had before. Pain persisted, but it had now turned either to resignation or hope for a better future. Some people felt that they are stuck in limbo, a situation of emergency that was not going to be solved, and thought about their future in pessimistic terms. Marcella, an old woman from L'Aquila, received a letter from the government telling her to leave her house and move to a hotel on the coast. She was sad and worried for the future and, even if she recognized the pain she experienced in L'Aquila, her roots were there and she did not want to leave her house and her city. That is why she asked for a form of assurance that she would be back by Christmas.

> It's sad to leave the tent and this city, actually to leave the city hurts. But when they move us I need the assurance to have a home by Christmas. I want a written commitment because today's letter has no legal value. I want in writing 'Marcella, you'll be home by this day, not your home, but a temporary one' and without this commitment I won't move from here.

Other citizens expressed a more positive attitude towards the future and already imagined the life they are going to live after this phase. For those involved in the earthquake, the pain of the loss of the lives they had before had been assimilated and metabolised, but it still surfaced at times. People remembered in front of the camera the lives they had before and the things that they had lost and that they missed from the past. For example, Angela talked about how her life had changed after the tragedy and about her new priorities:

> I was a lawyer. I still am, but I don't have a law office now. My office was close to centre and it collapsed. Our court of law was damaged and it's closed so until the end of the year there won't be trials. So my law office is in my tent. I put my computer on my ironing board and all my clothes as well are inside the tent. Of course working like this is almost impossible. I set some priorities for my life: get my kids out of the tent camp. So house first, law office after!

The pain in having lost her house with her precious things inside and for the deaths of many people is evident from this quote taken from a video about Nicoletta. But we can also find the will to move on, especially when the character underlined the good things that can come out of disaster, such as people helping each other to rebuild a better future.

> My house is destroyed too. I will never go there again. I'm not saying that I'm living well. I lost the house I love the most and I cried for two full days [...]. When you think of your plants dying, books getting wet [...]. I was able to cover the books, then I felt better! I started thinking about all the cards, the papers on my fridge, on the door [...]. All these things became piles of nothing. I lost those things. But there were some positive aspects too. I am sorry. I'm not ignoring the deaths, the losses and all terrible things we saw. I haven't forgotten the first two days I stood in front of the student campus because my friend had her son under the ruins there. We stayed there waiting to see if he was alive or not. I think I've seen the worst. It's hell to see parents hoping for their son to be alive then hoping their son died immediately so he didn't suffer it's the most terrible thing. I saw the worst, and I know it's there. But in tragedy and destruction some good came out, because people share and help each other. We had the energy of survival. It was a domino effect! So we had this energy and needed to do something good even though surrounded by destruction.

From the Interviews

From the thematic analysis of the interviews, we clustered the emerging topics into two main macro-categories. The first is related to the ability of FromZero TV to represent the daily lives of the people in the campsite and the second is about providing a faithful picture of the pain dimension. According to the majority of the interviewees, by telling their stories, the platform is able to give a truthful portrayal of the L'Aquila citizens' situation. FromZero TV is seen to describe convincingly the lives of the camp inhabitants who were forced to live together for several months and share, on a daily basis, both painful and hilarious moments with people they did not even know before the tragedy. The majority of the interviewees identify themselves with the characters of the videos and think that this television is able to go beyond the news, digging into the everyday reality of people facing emergency and building from destruction. In the words of Gino (37 years):

> I see myself in these videos. And I somehow come back with my mind to those days, those days of pain, those days of tears, when we were struggling for things that we used to take for granted and then were so hard to find [...]. Our everyday lives changed completely and we were living day by day, with not many expectations about the future. So yes, I recognize myself in these people and in their stories.

Also, Laura (28 years) provided similar reflections:

> That was actually our daily life in the tent camps, these videos made me think about all those hard moments and come back to the days I was with my family and with all those other people there for many months. Even if the videos are only showing a few stories, I think that they represent the core of our experiences that is the small things we have to face, all the time.

The second category refers to the representation of the pain of the victims. Interviewees have underlined that grief is portrayed by the videos of the FromZero TV television "with less tears and lights, not like a television show, but putting importance on our own views and our daily reflections and choices" (Laura, 28 years), "listening to our voices" (Mario, 42 years) and "respecting our feelings and above all, our pain" (Antonio, 21 years). Interviewees have highlighted that the pain was not turned into a media spectacle, nor "pumped" and that people were able to express their feelings in their own terms. Moreover, some interviewees pointed out that this Web TV not only shows the grief immediately after the tragedy on which traditional television has focused the most, but also the pain of the reconstruction. This point emerges from the words of Marina (31 years):

> These videos show people talking about what they have lost and what they want to do with their existences [...]. Those people are us [...]. We have lost so many things under those fallen buildings, we cried and we suffered, our pain is still here, even if the other media have gone, we are still struggling to go back to our normality and yes, these people have done a great job in showing this.

The ways in which this Web TV represented the pain of the victims was seen as "more nuanced than the one on traditional television" (Marina, 31 years), "full of the different aspects, not only negative, that life in the tent camps included" (Mario, 42 years) and "naturally flowing as a part of their lives, not as something useful for the media and then rapidly forgotten" (Chiara, 23 years). In the words of Barbara (38 years):

> It was a shock, indeed. We were all destroyed by the quake. Many people, friends, kids, families died. But there wasn't only tears and blood. Well, maybe at the beginning, in the first week. But then we moved on and we are still here. These videos show that we have learned

to live with that pain, that we want to start a new life, in our beloved city. There's a moment to just cry for days, nothing more and there's a moment to get up and recognize that you're lucky to have survived and that there are people like you around.

Final Remarks

In this paper, we have shown, through a series of examples, that traditional Italian television framed the post-earthquake situation as the "miracle in L'Aquila" and made a media spectacle out of the pain of those involved in the earthquake to impress the audience without respect for the feelings of the victims. Moreover, we have pointed out that it was on the Web that the victims of the tragedy could speak with their own voices and tell their own stories without filters.

Accordingly, our investigation, using qualitative methodologies (video analysis and semi-structured interviews), explored how the dimension of pain was represented on a cross-media platform that documented the lives of the people living in the tent camps after the tragedy.

Our hypothesis was that the platform's documentary style, with a strong focus on survivors' stories, together with the possibility of this web platform to bypass traditional gatekeepers, contributed to offer a coherent and more nuanced representation of the pain of those involved in the earthquake "on their own terms": closer to their everyday reality and less likely to be min into a TV Spectacular.

Findings from our research have confirmed our initial hypothesis. On the one hand, the analysis of the videos showed that the FromZero TV television offered a sober and even-handed representation of the dimension of pain in relation to the victims of the catastrophe. In contrast to traditional television, the FromZero TV episodes did not treat the dimension of pain in such an exposed way and did not make a spectacle out of it, instead choosing to respect the feelings of the people. On the other hand, the findings that emerged from the interviews confirmed that by giving a voice to the victims, this platform was able to represent grief in a more respectful way, adopting the words and the terms of the affected population without the biases that characterized traditional television.

Thus, our findings suggest that FromZero TV represents an interesting and powerful "experiment" of how new forms of television on the Internet can offer an alternative representation of events whilst giving a voice to ordinary people without having to appeal to the exhibitionism of feelings, of the media spectacle of catastrophe orchestrated by mainstream media.

Nevertheless, these new digitalized forms of television production cannot aspire to replace traditional television. In Italy, traditional television still represents the most viewed medium. In 2009, according to the CENSIS, 91.7% of Italian people accessed the traditional national television at least once per week compared with 15.2% accessing Web TV.

Thus, even if FromZero TV had provided an alternative view on the L'Aquila quake, it suffered from many of the same limitations that alternative media in general had to face in the case of the post-quake coverage: fragmentation (Fuchs, 2010), lack of consistent financial resources and lack of visibility when compared to the large numbers of the Berlusconi media empire (Farinosi and Treré, 2010a). As Couldry has recently pointed out:

> television may well remain the primary medium for most people for the foreseeable future, even if television content is for some audience sectors more often delivered via computers than television sets (Couldry, 2009, p. 443).

Further research is needed to explore the audiences of these new forms of television and to provide a richer understanding of the people who watch and benefit from this kind of content.

Moreover, we should not forget that, on the Web, the problem of public visibility persists because "also on the Internet political and financial power are essential for gaining public visibility" (Sandoval, 2009, p. 7). As Napoli (1998, 2008) has warned the Internet could undergo a possible "massification", where online activity will eventually converge around a few big sites which act like "traditional" gatekeepers. This would, in turn, obviate the opportunity for Web TV such as described in this chapter to be easily accessible or influential.

References

Barile L. 2009. Terremoto: un'informazione senza gente, senza volontariato. Reti solidali, 6 (2), pp. 16-18.

Bennett W. 2003. New Media Power: The Internet and Global Activism. In: N. Couldry & J. Curran (eds.). Contesting media power: Alternative Media in a Networked World. Lanham, MD: Rowman and Littlefield, pp. 17-37.

Boin A.,'t, Hart P., McConnell A. 2009. Crisis Exploitation: Political and Policy Impacts of Framing Contests. Journal of European Public Policy, 16 (1), pp. 81-106.

Bonaccorsi M., Nalbone D., Venti A. 2010. Cricca Economy. Dall A`quila alla B2, gli affari del capitalismo dei disastri. Roma: Edizioni Alegre.

Burkhart F. 1991. Media, Emergency Warnings, and Citizen Response. Boulder, CO: Westview Press.

Chomsky N. 1998. The Common Good. Tucson, AZ: Odonian.

Ciccozzi A. 2010a. Ad Reprimendam Audaciam Aquilanorum: cronaca e analisi del processo intentato da Bruno Vespa contro il dissenso aquilano. URL: http://lacittanascosta.blogspot.com/2010/04/ad-reprimendam-audaciam-aquilanorum.html (accessed 20 July 2011).

Ciccozzi A. 2010b. Dal sensazionalismo miracolistico agli stereotipi neorazzisti. URL: http://lacittanascosta.blogspot.com/2010/07/dal-sensazionalismo-miracol -agli.html (accessed 20 July 2011).
Colombo F. (ed.). 2004. TV and Interactivity in Europe: Mythologies, Theoretical Perspectives, Real Experiences. Milano: Vita e Pensiero.
Couldry N. 2009. Does 'the Media' have a Future? European Journal of Communication, 24 (4), pp. 437-449.
Dayan D., Katz E. 1992. Media Events. The Live Broadcasting of History. Cambridge: Harvard University Press.
Dominici P. 2010. La società dell'irresponsabilità. L'Aquila, la carta stampata, i "nuovi" rischi, le scienze sociali. Milano: Franco Angeli.
Farinosi M., Treré E. 2010a. Inside the 'People of the Wheelbarrows': Participation between Online and Offline Dimension in the Post-quake Social Movement. The Journal of Community Informatics, 6 (3). URL: http://ci-journal.net/ index.php/ciej/article/view/761/639 (accessed 20 July 2011).
Farinosi M., Treré E. 2010b. The Alternative Quake: Discrepancies between Mainstream and User Generated Information in the L'Aquila Case. In: L. Stillman & R. Gomez (eds.). Vision and Reality in Community Informatics. CIRN-DIAC Conference Proceedings: Prato, Italy 27-29 October 2010. CCNR Monash University, Information School, University of Washington.
Flick U. 1998. An Introduction to Qualitative Research. London: Sage Publications.
Fontana A., Frey J. 2005. The Interview: From Neutral Stance to Political Involvement. In: K. Denzin & S. Lincoln (eds.). The Sage Handbook of Qualitative Research, Vol. 3. London: Sage Publications, pp. 695-727.
Fortunati L. 2005. The Mediatization of the Net and the Internetization of the Mass Media. Gazette. The International Journal for Communication Studies, 67 (1), pp. 27-44.
Fortunati L. 2008. Mobile Convergence. In: K. Nyiri (ed.). Integration and Ubiquity. Towards a Philosophy of Telecommunications Convergence. Wien: Passagen Verlag, pp. 221-228.
Fuchs C. 2010. Alternative Media as Critical Media. European Journal of Social Theory, 13 (2), pp. 173-192.
Garnett J. L., Kouzmin A. 2007. Communicating throughout Katrina: Competing and Complementary Conceptual Lenses on Crisis Communication. Public Administration Review, 67 (s1), pp. 171-188.
Grasso A., Scaglioni M. (eds.) 2010. Televisione Convergente. La tv oltre il piccolo schermo. Roma: RTI-Reti Televisive It. Editore.
Imperiale A. J. 2010. Aftershock Communication. In: L. Stillman & R. Gomez (eds.). Vision and Reality in Community Informatics. CIRN-DIAC Conference Proceedings: Prato, Italy 27-29 October 2010. CCNR Monash University, Information School, University of Washington.

Jenkins H. 2006. Convergence Culture: Where Old and New Media Collide. New York: New York University Press.
Katz E., Liebes T. 2007. 'No more peace!': How Disaster, Terror and War Have Upstaged Media Events. International Journal of Communication, 1, pp. 157-166.
Kellner D. 2010. Media Spectacle and Media Events: Some Critical Reflections. In: N. Couldry, A. Hepp & F. Krotz (eds.). Media Events in a Global Age. London et al.: Routledge.
Lavanco G. (ed.). 2003. Psicologia dei disastri. Comunità e globalizzazione della paura. Milano: Franco Angeli.
Liebes T. 1998. Television's Disaster Marathons: A Danger to Democratic Processes? In: T. Liebes & J. Curran (eds.) Media, Ritual and Identity. London and New York: Routledge.
Lombardi M. 2005. Comunicare nell'emergenza. Milano: Vita e Pensiero.
Macias W., Hilyard K., Freimuth V. 2009. Blog Functions as Risk and Crisis Communication During Hurricane Katrina. Journal of Computer-Mediated Communication, 15 (1), pp. 1-31.
Masel-Walters L, Wilkins L., Walters T. (eds.). 1993. Bad Tidings: Communication and Catastrophe. Mahwah, NJ: Lawrence Erlbaum Associates Inc.
Meier P., Munro R. 2010. The Unprecedented Role of SMS in Disaster Response: Learning from Haiti. SAIS Review, 30 (2), pp. 91-103.
Micalizzi A. 2010. Memory of the Events and Events of the Memory: How the Net can Change our Way of Constructing Collective Memory. In: Proceedings of ESA Research Network Sociology of Culture Midterm Conference: Culture and the Making of Worlds, October 2010. URL: http://papers.ssrn.com/sol3/papers.cfm?abstract_id=1693111 (accessed 20 July 2011).
Miller A., Goidel R. 2009. News Organizations and Information Gathering During a Natural Disaster: Lessons from Hurricane Katrina. Journal of Contingencies and Crisis Management, 17 (4), pp. 266-273.
Morcellini M. (ed.). 2006. Torri crollanti. Comunicazione, media e nuovi terrorismi dopo l'11 Settembre. Milano: Franco Angeli.
Muralidharan S., Rasmussen L., Patterson D., Shin J. H. 2011. Hope for Haiti: An Analysis of Facebook and Twitter Usage During the Earthquake Relief Efforts. Public Relations Review, 37 (2), pp. 175-177.
Napoli P. 1998. The Internet and the Forces of "Massification". The Electronic Journal of Communication, 8 (2). URL: http://www.cios.org/ejcpublic/008/2/00828.html (accessed 20 July 2011).
Napoli P. 2008. Hyperlinking and the Forces of "Massification". In: J. Turow & L. Tsui (eds.). The Hyperlinked Society: Questioning Connections in the Digital Age. Ann Arbor: University of Michigan Press, pp. 56-69.
Padovani C. 2010. Citizens Communication and the 2009 G8 Summit in L'Aquila, Italy. International Journal of Communication, 4, pp. 416-439.

Pasquarè F., Pozzetti M. 2007. Geological Hazards, Disasters and the Media: The Italian Case Study. Quaternary International, Vol. 173-174, pp. 166-171.
Patton M. 2002. Qualitative Research and Evaluation Methods. London: Sage Publications.
Perez-Lugo M. 2001. The Mass Media and Disaster Awareness in Puerto Rico. Organization & Environment, 14 (1), pp. 55-73.
Perez-Lugo M. 2004. Media Uses in Disaster Situations: A New Focus on the Impact Phase. Sociological inquiry, 74 (2), pp. 210-225.
Ploughman P. 1997. Disasters, the Media and Social Structures: A Typology of Credibility Hierarchy Persistence Based on Newspaper Coverage of the Love Canal and Six Other Disasters. Disasters, 21 (2), pp. 118-137.
Puliafito A. 2010. Protezione civile Spa. Quando la gestione dell'emergenza si fa business. Reggio Emilia: Aliberti.
Quarantelli E.L. 1991. Lessons from Research: Findings on Mass Communication System Behavior in the Pre, Trans, and Postimpact Periods of Disasters. Preliminary Paper No. 160. University of Delaware, Disaster Research Center. URL: http://dspace.udel.edu:8080/dspace/bitstream/handle/19716/532/pp160.pdf (accessed 20 July 2011)
Rucht D. 2004. The Quadruple A: Media Strategies of Protest Movements since the 1960s. In: W.B. Van De Donk, B. D. Loader, P.G. Nixon & D. Rucht (eds.). Cyberprotest. New Media, Citizens and Social Movements. New York: Routledge, pp. 29-56.
Sandoval M. 2009. A Critical Contribution to the Foundations of Alternative Media Studies. Kurgu-Online International Journal of Communication Studies, 1, pp. 1-18.
Scaglioni M., Sfardini A. 2008. Multi TV. Roma: Carocci.
Scannell P. 1996. Radio, Television and Modern Life. Cambridge: Blackwell.
Scurati, A. 2003. Guerra. Narrazioni e culture nella tradizione occidentale. Roma: Donzelli.
Skinner J. 2004. Before the Volcano: Reverberations of Identity on Montserrat. Kingston: Arawak Publications.
Strocchi S. 2009. Filmmaker Q&A: From Zero. When the International Media Left an Italian Disaster Area, Italian filmmakers Stepped In. URL: http://english.aljazeera.net/programmes/witness/2009/12/2009121911952220509.html (accessed 20 July 2011).
Tota A. 2003. La città ferita. Memoria e comunicazione pubblica della strage di Bologna, 2 agosto 1980. Bologna: Il Mulino.
Treré E. 2008. Problematiche e prospettive delle Web TV universitarie italiane. In: G. P. Caprettini & L. Denicolai (eds.). Extracampus. La televisione universitaria. Case-history di un'esperienza vincente. Torino: Cartman.

Treré E., Bazzarin V. 2011. Exploring Italian Micro Web TVs: how high-Tech bricoleur redefine audiences? ESSACHESS Journal for Communication Studies, 4, pp. 49-67.
Treré E., Sapio B. 2008. DTV in Italy. In: W. Van der Broeck & J. Pierson (eds.). Digital Television in Europe. Brussels: VUBpress.
Turner R., Nigg J., Paz D. 1986. Waiting for Disaster. Berkeley, CA.: University of California Press.
Vincent J., Fortunati L. (eds.) 2009. Electronic Emotion. The Mediation of Emotion via Information and Communication Technologies. Oxford: Peter Lang.
Yates D., Paquette S. 2010. Emergency Knowledge Management and Social Media Technologies: A Case Study of the 2010 Haitian Earthquake. International Journal of Information Management, 31, pp. 6-13.
Walters L. M., Hornig S. 1993. Faces in the News: Network Television News Coverage of Hurricane Hugo and the Loma Prieta Earthquake. Journal of Broadcasting & Electronic Media, 37, pp. 219-232.

Part II

Digital Television Audiences and their Practices of Use

Leif Kramp

Access to Cornucopia? The Rise of a New Television Archive Culture on the Web

Introduction

The transformation of media production and use through the growth of the Internet has increasingly caused shifting paradigms among the established media. There is much debate within academia as well as the industry on whether television, in its traditional forms of distribution and reception, has a future or whether it will be subject to radical changes resulting from digital retrieval principles and interaction potential of the Internet. Silverstone sees media and society at a tipping point in today's thoroughly mediatised environment where a "mediapolis" exists

> both at national and global levels, and where the materiality of the world is constructed through (principally) electronically communicated public speech and action (Silverstone, 2007, p. 31).

In this context, audiences break out of their dependency relationship with the formerly all-powerful television networks when it comes to questions of supply and demand, but without abandoning televisual content. The Internet enables an increasingly autonomous fusion of media consumption, production, and distribution on the users' side.

Focussing on the shifting role and power relations between media users and producers, this chapter explores the implications for access to formerly aired (that means historical) television programmes. The broadcasting archives of the television industry are an institution that primarily fulfils a functional task by supporting television production, filling gaps in the programming schedule, and generating profit for the entity of which they are a part. These archives are, by virtue of their traditional constitution as departments of commercial enterprises, neither easily accessible nor monitored transparently by the public. Contemporary history, the *zeitgeist*, and the memories of generations connected to the moving images that were circulated through the "goggle-box" depended on the fluctuating and fragmentary rerun offerings of the programme flow or the sporadic release of singular programmes on DVD.

> Effective democratisation can always be measured by this essential criterion: the participation in and the access of the archive, its constitution, and its interpretation (Derrida, 1996, p. 4).

Following Derrida's constitutive remarks on the relevance of archives for the democratic social order, public access to, and functionalisation of the broad and diverse heritage of television appears to be a basic requirement in Silverstone's concept of a "mediapolis".[1]

Archives play a central role as a guidance aid within social identity work (Blouin and Rosenberg, 2007). The more complex and fragmented a society is, the more important are archival services that take these complexities, fragmentations, and consequent insecurities into account. Therefore, the archival practice cannot be devoted only to administrative processes or to a one-sided productive goal, they must also acknowledge the needs of a pluralistic society. For that reason, the German historian Winfried Schulze claimed that archives in all provenances have inadequately understood themselves as having a broader duty to ensure that cultural heritage is accessible for social reminiscence work on a much wider scale (Schulze, 2000, p. 27). The symptoms of the limited accessibility of archived assets, especially in the television industry, lie mainly with the self-management responsibility and the consequential access regulations of the institution that controls access to its archive and collections.

To determine whether the established access structure might undergo a reconfiguration under the influence of user activism on the Internet, it is critical to examine how archival authorities evaluate the engagement of the audiences who persistently try to access television content, regardless of the programming schedule and conventional sale offers. This is coupled with the proliferation of audiovisual and specifically televisual content available to audiences from the digital media sphere via *YouTube*, peer-to-peer-networks, and various knowledge bases populated by fans and television buffs. The following considerations are based upon an empirical survey among institutions that belong to the emerging field of television and audiovisual heritage management. The survey consists of 55 semi-structured interviews with 63 experts from the television industry, the public sector, and academia in the United States, Canada, and Germany, which were conducted between 2006 and 2009. To gain a comprehensive picture of the work of television-related archives, libraries, and museums, the sample includes 15 broadcasting archives and three university archives, two libraries, a mediathèque, and a mediathèque network, as well as ten museums. Where possible, more than one responsible representative of a major institution was interviewed to obtain a com-

1 This was also recognised in part by supranational organisations such as UNESCO in the "Recommendation for the Safeguarding and Preservation of Moving Images" (1980) and the Council of Europe in the "Protocol to the European Convention for the Protection of the Audiovisual Heritage, on the Protection of Television Productions" (2001).

prehensive picture of appraisal and experience. Additionally, a selection of independent experts from academia and heritage management was consulted. These experts were selected based upon their respective publication record, international expertise, and experience in archival and museum practice, theory, and televisual memory work.[2]

From Collecting to Sharing Television Heritage

If the remote control has already been declared as the incisive instrument of a "television revolution" in the living room, it was actually the video recorder which changed the relationship between the viewer and television more profoundly. As an "audiovisual time machine", (Zielinski, 1999, p. 236) the television recording device put the user in a position where for the first time they were able to free themselves from the determination of the linear reception within the boundaries of the programme structure. Access to TV content was no longer exclusively determined by the broadcasting schedule but rather it fit into the daily routines of the viewer much more comfortably.

In recent years, the rapid expansion of the video-on-demand (VoD) practice on the Internet complemented the analogue storage techniques:

> The Internet has served the video collector well in building trading networks and fan communities, yet it ultimately runs counter to the TV collector's chief aim: to make the immaterial material (Bjarkman, 2004, p. 240).

Going even further than analogue and digital recording devices, VoD suspends the "central organisation of circular effects" (Zielinski, 1999, p. 239) as an unusual form of distribution ("publication principle" vs. "programme flow"; Miège, 1989, p. 12).

---

2 The sample included the archives of the US networks ABC, CBS, CNN, FOX, NBC, PBS, the Canadian CBC, and the German broadcasters ARD (DRA), BR, NDR, ProSiebenSat.1, RTL, SWR, WDR, and ZDF. Further surveyed archives, libraries, and museums were the UCLA Film & Television Archive, the Vanderbilt Television News Archive, the Walter J. Brown Media Archives and Peabody Awards Collection, the Library of American Broadcasting, the Library of Congress/NAVCC, the Netzwerk Mediatheken, the CBC Museum, the Museum of Broadcast Communications, the American Museum of the Moving Image, the MZTV Museum, the Newseum, the Paley Center for Media, the German Kinemathek Film and TV Museum, the German Museum of Technology, and the Museum of the History of the Federal Republic of Germany. Additional independent experts came from the Massachusetts Institute of Technology, the New York University, the Northwestern University, Syracuse University, the University of Georgia, the University of Toronto, the University of Wisconsin, and the Humboldt University of Berlin.

The recording, storage and retrieval technologies have created a new way to watch television as the viewer is liberated from the principles of the programme schedule with its fixed broadcast dates because it is now possible to watch one's favourite TV shows anywhere, at any time. With VoD, the TV industry itself initiated the end of the old television broadcasting age. The progressive integration of marketing principles leads to a supplementation of their respective underlying objectives and can cause an increase in mnemic qualities of television programmes and their contents. The loyalty of the widest possible audience to a broadcaster's brand ("flow") and the sales maximisation of a product unit by the increase in attraction of certain items ("publishing") are alike in their efforts to win users over to an abstract media product (programme line-up in its entirety) or a specific one (distinct programme/production). Thus on the one hand, awareness of the broadcaster's brand as a whole is generated more intensively, while at the same time some single formats are also advertised. These strategies combined may potentially lead to higher levels of the consumer-product relationship and will potentially increase the likelihood that users are able to remember television content and its context. With the retrieval options available day or night, and with the powerful recording tools available, a deliberate selection process to watch a specific programme becomes possible.

However, the parallel offering structure of content retrieval through the VoD services of the industry is complemented by an anarchic movement on the users side that harnesses the powerful recording tools of the Web to collect, share, discuss and cultivate TV content – which all has a radical impact on access to the wide television heritage. Television viewers are plundering their analog video collections and scouring the TV schedules for reruns to fill the insatiable Web reservoirs with programming from yesteryear. Concomitantly, the volume of audiovisual media content that is permanently available for reception is continuously growing. Rapidly increasing – and affordable – storage capacities for digital media have led to a worldwide explosion of freely available data.

YouTube has played a significant contribution in the metamorphosis of the recipient becoming a "prosumer" – a producer *as well as* consumer – of audio-visual works.[3] The available content is not limited to user-generated videos: with the strong desire to experiment, and all too often in disregard of statutory provisions, users will record, copy, edit and publish television material from wherever they can get hold of it. Whilst *YouTube* is the leading worldwide video portal, with over two billion views a day, it is certainly not the only online platform that hosts records and copies of television material. Numerous providers try to profit from the activity of their users – "the people formerly known as the audience" (Rosen, 2006) – and also offer the option to upload video content for free in order to enhance the usage numbers. In addition, so-called peer-2-peer or file-sharing net-

3 For a critical discussion of the term "prosumer" (Van Dijck, 2007).

works such as *BitTorrent*, *Gnutella* or *eDonkey* are used to share digitised movies and TV shows in full length and in high quality. This results in an aberrant, often personalised, juggling act with television content following a dominant sharing principle instead of the organised mechanisms of linear supply.

Archival Access at the Crossroads

The rigorous change in television use has not yet led to general access to the digitised archival assets in the TV archives though. However, it is highly likely that the activity of the audience, which shows no fear of resurrecting television heritage on their own, could trigger a further rethinking process by the television networks. After all, *YouTube* has not become the world's most infamous and frequented video portal because TV producers were so heavily engaged: it was rather the user's achievement to create a new archive culture on the Web as a by-product of social activities that is constantly changing in form and stock (Gracy, 2007, p. 196).

Access to old television programming is no longer solely controlled and determined by institutional efforts of collection, preservation and distribution, but also by acts of copyright infringement. As the survey of major institutions belonging to the television heritage management in North America and Germany has shown, television heritage is mainly locked up behind the corporate walls due to commercial interests and rights issues (see Table 1).

While the traditional market cycles are still in place, broadcasters and producers are increasingly having to adapt to the collaborative and communal spirit of the Internet. It enables optimised searchability, linking and accessibility of data repositories and proves to be an ideal complement to the existing institutional archival services. More and more users are putting their television recordings online, and it does not end with video material; it also includes context information, trivia and memories. Therefore, TV viewers who are active on the Web have to a certain degree emancipated themselves from the institutional access regulations of the TV sector, sharing their own collections and resources with each other.

The most sought-after target groups of under 29-year-olds tend to be far from willing to access the television heritage by traditional means:

> We live in a visual universe and the thought that the content resides at the Library of Congress or somewhere else and they have to go there: Forget about it,

says Jane Johnson, project manager of the joint online-portal *Moving Image Collections* (MIC) (cited in Kramp, 2011, p. 275). User expectations are now obviously so high that access to archive materials is subject to a completely different set of determinants than before. Neither television broadcasting operations nor the

	Discovery	Viewing	Reproduction	Use (rights clearance)
Television broadcast networks' archives, both public and commercial	Networks do not typically reference footage other than their own. Research services for private and academic use are usually not provided.	Varies widely by network, heavily restrictive, but there is a trend toward online viewing.	Networks usually provide reproductions of news where all rights are with the broadcaster, but don't always own and thus can't reproduce entertainment footage.	Networks sell usage rights to their news, but do not always own (and thus cannot clear) entertainment footage. Third-party rights cannot be negotiated.
Commercial providers and monitoring companies	Commercial sources are useful for advertisements and some news; less useful for entertainment footage that is not for sale.	Higher costs, but generally fast response times. Viewing only after fee required ordering.	Reproductions are available for purchase.	Commercial providers can handle rights clearances.
University media libraries	Only a few university libraries have substantial video collections. Often heavily fragmentary. Research in collections only on site.	May require travel or in exceptional cases ordering of videotapes by mail (news programming). Access on site mostly restricted to university members or visiting fellows.	Concerns about potential liability cause university libraries usually to restrict access to and copying of video footage, though news footage can be loaned.	University libraries may provide limited assistance in exceptional cases.
Public institutions: special collections, libraries, and museums	Access to video broadcast on multiple networks, but may have less comprehensive holdings than broadcast networks. Collections are easy to discover.	Does require travel. Unrestricted access.	These organisations must carefully abide by the restrictions placed on them by owners. Usually no reproduction of archival holdings.	These organisations may provide limited assistance.
Fan clubs/private collections	Coverage is spotty, difficult to locate and to research.	Inconsistent. Depending on willingness of the collector.	Reproductions are easy and convenient but legally generally problematic.	Rights clearances by these groups/collectors unlikely.

Table 1: Status quo of access to television heritage. Source: Ubois, 2005 and own survey.

home video market determine how this content is used in the Web 2.0 era, but it is actually the collaborative as well as communal spirit of the network under whose auspices television collections are pushed out from the obscurity of the attic into the public spotlight. Internet users are not obviously (only) dependent on the repetition of programmes, on DVD releases, even recording services, or the serendipity of the television archivists to come across historical television works. They have become emancipated – to a certain degree – from institutional access regulations by supplying each other with programme material. The determinants of audience participation in public discourse on television (and with television material) have changed from the ground up:

- The rapid technological progress of the Internet has led to a lively *exchange of current and past programming* (audio-visual works as well as context information). In the analogue media environment, it was only practical for audience members at considerable expense to share their own private collection of video cassettes with other viewers such as fans who have the same likes and dislikes. This required considerable commitment and sacrifice of individuals to copy cassettes with the aid of two video recorders and to send the duplicates by mail to the interested parties. Digital tools have simplified this process significantly.
- With the increasing use of the Internet by wide sections of the audience, not only has the capacity of networks been expanded, but also the *development of interactive publishing and communication tools*. One consequence has been the formation of social networks that enable the user to link their contents mutually and to implement powerful recommendation structures that prepare the recipients to watch and distribute television content their way. The willingness to share one's own television experiences with other users is therefore not only a result of the enhanced technical possibilities, but also and above all the result of the optimised social interaction.
- Grassroots movements are characterised by a pioneering spirit, without which a self-conscious appropriation and cultivation of innovative forms of interaction and discourse by a broad user base would be unlikely. The communicative and informative variety of the Internet results from a newly arisen "no-cost" mentality, which suggests to the user that the content overload of the deregulated global network is, per se, retrievable at no charge (Whyte, 2001, p. 48; Albarran, 2010, pp. 137f). This kind of "gratis-mania" (Schmalz, 2009), that turned consumer psychology upside down, has of course serious implications for the *general sense of wrongdoing in the use, publication and sharing of media works*.
- The almost unmanageable amount of freely accessible content on the Internet creates a sense of limitless freedom in the virtual exchange of material. It also, potentially, allows all users *to put together a special online collection of*

> *television content from the depths of the network and contribute their own notes or copies.* The strong sense of community encourages the users to make use of available third-party content and to utilise it for their own purposes. The willingness to use television material for autobiographical reminiscence work and to express their individual identity in terms of defining television experiences appears to have spread significantly.

The variety of audio-visual material that resides online is now so beguiling, that one could already have the impression of a cornucopian Web inventory of the media heritage:

> Have you noticed that kids – and many adults, too – think every article ever written and every song ever sung is on the Internet? It will not be long now before young people will grow up assuming that every TV program ever made is online, too. That's what they will expect (Rubin, 2007).

However, the assumption of broad availability is clearly illusive since large parts of the archival holdings have not yet been digitised. As Chuck Howell, curator at the Library of American Broadcasting, notes, the Internet only seems to be filled with exhaustive archival resources. However, research into TV's past on the Web could only be a cursory search. With the legal barriers and related restrictions, no scholar could get past the traditional way of research, i.e. to visit an archive personally and incorporate oneself locally in the stock material (see Howell, 2006, p. 305). So as promising as the virtual corpus presents itself to be as a source to provide quick answers to fundamental questions, its basic quality is not the holistic curation and long-term preservation of a repository, but rather what the television clips are actually *used* for by the audience.

Remembering (with) TV against All Odds

The significance of the grassroots activities on the Web lies consequently in the unorthodox and eclectic usurpation and mastery of television's past by the users. This archival concept is no longer based on the institutional collection mission, controlled preservation efforts and restrictive access, but on the collectivised and creative functionalisation of single programming excerpts ("Clip-Culture"; Anon, 2006). Here, the goal is not to preserve the original in its best possible condition, but to edit, annotate and circulate television content and communicate about it in the "here and now" through different instruments such as Web portals, digital databases, communication tools, social networks, etc. (Burgess & Green, 2009, pp. 68f). The focus therefore lies on the motive to get not just mere access to television's past, but to use it actively and creatively to achieve personal communication, socialisation and mediation goals. However, continuity in the online availability of the used programme material can be guaranteed in the rarest of cases because the users do not have ultimate control over their digitised collections that are predominantly hosted by external Internet companies that are prone to delete content if evidence occurs that its publication is part of an infringement such as copyright violation.

Nevertheless, users will search and find new ways to get hold of the desired material, or even offer it, although it will mean a violation of statutory provisions. This revolt against the established law turns the inventive and access-seeking viewers in the network into "cultural outlaws" (Costello and Moore, 2007) who are pushed into the roles of activists due to the criminalisation of their desire for free access to digitised television heritage. The revolutionary formula of the new archive culture on the Web seems to be that if a television production is broadcast only once, it quickly finds its way into the public domain.

The fight over the hegemony of access between the industry on one side, represented by media producers, access providers and diverse rights owners, and the users on the other, has escalated several times, additionally because of criminal profiteers who offer content to stream and download illegally with a view to gain. Therefore, the free use of the television heritage is inhibited by so-called "copyright horror stories" which, although not primarily in the television business, have taken place in media-related markets (Boynton, 2004; Vaidhyanathan, 2004; Kirk, 2011; Walker, 2011). Billions in claims, as in the case of a dispute between the media conglomerate Viacom and Google-owned *YouTube* that attracted attention, convey an unmistakable message: unauthorized use of television material is not a trivial offence.

In addition, the measures taken by the media industry to undermine the upcoming setting up of a self-generating collective online archive of countless recordings and copies are proving to be increasingly effective: innovative search and filtering technologies with names like "Take Down Stay Down" (Lowry, 2007) allow operators of sites like *MySpace* and *YouTube* to detect and suppress the unauthorized publication of copyrighted work or portions of it automatically, as long as the individual video file has already been registered once as a violation of the copyright.[4] Even if the rights owners impose rigid cancellations and closures of illegally retrievable programmes or programming clips, the content often appears quickly on another platform. For example, when *YouTube* deleted the nine-minute parody about the TV series "Twin Peaks" from the comedy show "Saturday Night

4 The force fields between the users and the media industry are not only characterized by a constant modification of power relations regarding the control of available content. On a higher level the issue is shaped by regulatory questions of broadband Internet services and network neutrality (Lenard and May, 2006) that have wide consequences for do-it-yourself archiving and access provision through the activities of the user collective, especially when it comes to storage capacity limits as well as collecting and accessing high-resolution files.

Live", the clip re-emerged a short time later in full on other video portals such as *Dailymotion.com.*

The permanent release and demise of televisual shreds is akin to the struggle between remembering and forgetting, as imponderable as human memory but also with surprising rediscoveries:

> Like memory (cultural or personal), YouTube is dynamic. It is an ever-changing clutter of stuff from the user's past, some of which disappears and some of which remains overlooked, while new material is constantly being accrued and new associations or (literally, hypertext) links are being made. The images are often hazy but may suffice to induce recall or to fill in where we could only previously imagine how things were from written or word-of-mouth accounts (Hilderbrand, 2007, pp. 50).

Hilderbrand understands *YouTube* as a technicistic and aesthetic concept of a new form of memory culture: the TV clips that are uploaded by users largely indicate the recording procedures and contexts, for example by the lower image and sound quality such as blurs and colour streaks but also through graphic insertions such as station logos and overlays. They therefore become a testimony of the targeted collection activities of recipients which deny the institutionalised access hegemony of the television industry and characterise themselves as a selection and distribution agency. *YouTube* evolved through this active participation of tens of thousands of amateur collectors not only into a hub for user-powered online video portals, but also into a synonym for an aesthetic of access to the cultural moving image heritage.

This kind of self-generating archive obeys neither an institutional control nor a clear business strategy and presents itself as a by-product of social activities and subject to constant change and progression (Gracy 2007, p. 196).

> Once people have got that material, it's really hard to stop them to put it on an illegitimate platform on the Internet,

says Robert Thompson, Trustee Professor of Television and Popular Culture at the S.I. Newhouse School of Public Communication (cited in Kramp, 2011, p. 276). Thompson explains the urge of many users to be a part of the pluralistic distribution and sharing network for television material. TV shows do not serve primarily as occasions for "water-cooler conversations" any more (Einav and Carey, 2009, p. 128), but they have become an illustrative and expressive tool for personal feelings and an object of discourse in the best possible sense. The viewing experience of those who watch programmes on the Internet is asynchronous and yet it is viewed as a communicatively knit activity of virtual communities and an intellectual exercise at the same time – the contents are collectively discussed, annotated, and sometimes even satirised as a form of media critique.

Collect, Share, Communitise

TV programmes in the form of clips, selected, compiled and commented on by viewers, may not be described as pure consumer product. They rather serve as "building blocks of creative acts or public speech acts" (Gracy, 2007, p. 183). In the personal context of collecting television programming, exhibiting and sharing the material is the focus but what often follows are emotional imperatives that connect to the content of the programming to express feelings and accentuate world views, just as with views on life and history. The basis for the informal use of professionally produced audio-visual media works is an altered social behaviour that has emerged under the influence of communicational networking and the equalised and volatile sender-receiver relationship on the Web. Jenkins suggests speaking preferably of "communal media" instead of "personal media" which are not isolated but permits a new access to the community and therefore are ideal for innovative forms of communitisation (Jenkins, 2006, p. 245). Thereby the contexts of remembering television are in a process of remodelling: programming content in its audio-visual concreteness offers reference points for virtual group formation. Thus after decades of absence in terms of a functional electronic hearth, television is used as just this, if only in the disembodied environment of the global network. Long before the digital media revolution, fan communities and interest groups have been formed around television myths, but with its numerous articulation opportunities and communication modes, the Internet turns out to be the ideal environment to build an "archivalesque" network as a combination of multiple memory communities that use the history of television as a unifying element.

First and foremost, fan groups show various efforts to experiment and participate with the interactive online tools. Thanks to the Internet technology, fan originated and managed content underwent an unprecedented proliferation and granted access to a wealth of archival material from every imaginable area of television history that were previously not publicly available, because hitherto individuals had only rare opportunities to present single collection items to a small audience on occasions like fan meetings. Fan websites can be committed to popular and worldwide known formats like "The Twilight Zone", which even had a virtual "museum" dedicated to it by passionate devotees, right up to individual, less well-known television personalities such as Lara Logan, senior foreign correspondent of CBS, whose reporting, as well as appearances on talk shows, are closely followed by her fans and presented on the Web.

In this way not only the fans, but also unspecified members of the public, can acquire a taste for the specific topic, be it a popular franchise like "Star Trek" or a person-centred fan culture. Fan activities are strong promoters of awareness for the use of the available sharing tools, because they do not address the general public on an abstract and random level but by direct confrontation with the audiovisual products to animate autobiographical memories about the individual's television past. The willingness to participate in this discourse in the form of text elaboration and by contributing gathered television clips or their own recordings is highly likely due to the ease of the encoding and publishing arrangements on the Web.

What is more, collectors have discovered the Internet for themselves. Extensive private collections were almost inaccessible in the analogue age of media and they were only accessible to people known and trusted by the individual collector, since any insight by external users required great effort and the disruption of privacy. Digital platforms provide the possibility to proudly present collections that were gathered in private and distinguish oneself as an "expert" for TV history purposes. On *YouTube*, numerous collectors exhibit excerpts of their recording repertoire that document the extent of their television archive. This approach comes with delicate rights issues because individual collectors will, due to the extraordinary efforts and troubles required to search for unknown rights holders, ignore the legal requirements to publish material and they will therefore publish without legal protection. A prominent example is the company Video Resources that is owned by the television writer and director Ira Gallen in New York. Gallen amassed a private collection of nearly 10,000 early commercials, variety shows and other programming relics from the "golden age" of US television and in 2007, he started to upload a significant amount of programming material for his project "TVdays" for free. His ambition was to call attention to his services: Video Resources licenses the historical material to companies that might be interested in reusing it. For each clip, Gallen charges between US$ 2,500 to US$ 3,000 but he does this without obtaining a permit from the hard to find rights holders or dissipating a portion of its sales to them (see Sarno, 2007).

Some devotees of television have also committed to the overarching goal of providing guidance in the current and historical diversity of programming. Ordinarily initiated by individual users, websites such as *TV Tome* (now *TV.com*) or *Sitcoms Online* evolved to encyclopaedic databases of television productions thanks to the participation of numerous users who contributed their detailed knowledge on actors and episodes or by simply supplying trivial anecdotes about the show. These websites often rise to the position of central reference points or even clearing houses as they provide details about other sources of information such as the availability of DVD versions of a historical programme or the opportunity to watch them on certain online portals. Although being a reservoir for all kinds of user interests, scholars are one of the real profiteers of this swarm of television intelligence, as Lynn Spigel, Professor of Screen Cultures at the School of Communication at Northwestern University, points out:

> Things like *TV.com* are very helpful: You can go online and find endless episode descriptions. That was not available in the early 80s (cited in Kramp, 2011, p. 278).

Although *TV.com* has been bought by CNET and subsequently been passed on to media giant CBS Interactive, there is a vivid scene of Web based information services that are primarily run by television enthusiasts. Even minor issues are addressed extensively in order to serve the knowledge zeal of the diversified fan communities. Thus, the independent researcher Jerome Holst attempts to list as many TV characters as possible on his website *TVAcres.com*, including those in the background, places, cars or other areas which appear on US and British television programmes screened on primetime and Saturday mornings.[5] It is therefore crucial, ultimately, to know how and why the young and old television heritage is used by the dedicated audience.

Through the Archival Barricades

The many different varieties of media available have already revolutionised the relationship between the audience and television history. The self-motivated urge of the recipient in making available selected programming excerpts, facts, trivia and memories has become a driving force in shaping public memory cultures and it is increasingly influencing the strategic actions of institutionalised television heritage management:

> YouTube's archive made the videos accessible, linkable, and discussable in a way that motivated further rhetorical activity and that transcended YouTube's structure even as it reinforced it (Skinnell, 2010).

Nevertheless, the barriers to access the historical richness of television are still not to be underestimated. The previous user initiatives could reveal the tips of a myriad of icebergs. The demands for an adjustment of traditional access models of existing conservation and heritage management facilities are high. Television content is expected to be cross-linked online and to be individually integrated in different publication and functionalisation contexts such as blogs, social networks or websites of virtual memory initiatives. The fact that the liberation process on the side of the user is limited by legal and economic conflicts, but also because access to television history through private collections of recordings and documents has inevitably been viewed as insufficient quantitatively, shows the randomness of such compartmentalised archive-structures.

---

5 Examples from Germany range from episode guides (*fernsehserien.de* with data of more than 13,700 German and foreign television series), to transcripts of specific episodes as on *TV Serien Infos* (*tvsi.de*), to the meticulous tracing of historical programming days that are reconstructed from old television magazines as on *tvprogramme.net*, or to websites that publish old letters to TV magazines as on *zuschauerpost.de*.

Although it is not an alternative for the protection of audio-visual heritage, it still gives a major stimulus to responsible archival entities in the media business and the cultural and education sectors to rearrange their modes of operation to provide the general public with access to television heritage by opening it up to audience members who would like to help. Clearly new demands on the handling of archived digital content have emerged. Television heritage management – whether it is for-profit within the industry framework or not-for-profit in terms of public service institutions, such as museums and libraries – can benefit from the publics' readiness to participate. As Rick Prelinger points out, archives have to rethink their canonical missions and proceedings:

> Like libraries, publishers and record labels, moving-image archives suffer from clinging to outdated paradigms of access and distribution. Change is urgent. But the developments that have weakened archives' status in the public sphere and threaten to subvert the consensus that keeps them alive aren't primarily consequences of their own actions or inactions. Rather, their loss of status is a result of powerful externalities over which they have no power (Prelinger, 2009, pp. 273-274).

Annotation projects with user involvement promise to be an exciting but also tricky way of engaging the uncontrollable movement on the Web. To introduce a complex archival inventory, you need powerful search and distribution mechanisms – and manpower. As the volume of to-be-archived programming increases, the pressure on cataloguing measures weighs heavily on the archives. An archivist needs up to seven hours to catalogue a 50-minute news broadcast (Dowman et al., 2005, p. 225). It is therefore not surprising that the gathering of archival metadata can quickly fall behind. With so-called "folksonomies" Internet users can become an active part in the indexing process by providing user-generated metadata. Archivists have already dealt for some time now with the question of whether social tagging, i.e. the annotation of programme material with key terms by the mass audience, is an effective alternative to the traditional ways of archival indexing (Trant, 2009). The downside concerns the restrictions, the unreliability and the lack of control options for user-generated metadata.

The average user, who is to be addressed with such a strategy, commonly contributes nothing either to the frame or core data, the condition of an archival record or the rights information. Their role is limited solely to the enrichment of content descriptions and contextual information. The high risk of errors in the description, whether of content or language, may limit the expected benefits. In addition, the enforcement of a standardised terminology due to the openness of the annotation procedure is unlikely to be possible. There is also the danger of the data volumes being too complex which can lead to an accumulation of irrelevant search results, which in turn can be adjusted only by means of editorial supervision by specialists. On the other hand, this very complexity can be utilised with the help of effective digital filtering tools and can create a greater number of access points into the depth of the archival assets compared to traditional cataloguing methods. It can also be assumed that some user groups have high levels of experience in the implementation of appropriate tools. The active participation of users on websites and social networks like *Wikipedia*, *Facebook*, and *YouTube* illustrates the growing motivation of the audience to become active participants in knowledge creation:

> User-created metadata is not a panacea for archival cataloguing woes, but it is an important strategy that should be considered seriously for its potential to provide cheap and easy content level descriptions that are catered to the user. Over time, and with different levels of editorial control, user-created metadata could prove very useful to the archival community (Andreano, 2007, p. 92).

Finally, the participation of recipients in the generation of descriptive data on historical television stations is of immense importance from the perspective of memory theory. Thereby they add their own interpretations, value judgments and memories. This also has economic implications as the audience can be targeted more personally and efficiently and can be connected more closely to archived television content that is marketed through cross-media supply chains.

Access to once aired television material is a key factor to arouse awareness and willingness of wider parts of the audience in order to devote their time, knowledge and experiences for such projects without weakening the central authority role of the professional heritage management of memory institutions. Public archives, libraries and museums that fulfil the task of managing the television heritage to preserve it in their traditional role as documentary institutions and advocates of access seem to become slightly obsolete at first glance when the audience as well as the television industry pave the way to historical television programmes in a dynamic interplay for, but also against each other. However, developments to date suggest that for a foreseeable future a reliable availability of a broad corpus of programming material from the history of television for general use on the Internet cannot be provided by the market nor by the users themselves.

The future of television heritage management cannot be developed collaboratively under the auspices of the anarchic and chaotic archival structures of the Internet, but has to follow the systemic environment of professional supervision by the institutionalised centres of excellence that are the broadcasters and public media libraries. This also means, however, a strict commitment to a qualitative improvement of accessibility standards in order to increase the use and value of archival assets that may in turn evoke a continuously growing demand for access by the audience and launch a more strategic opening of the archives. Respectively, Table 2 shows – following the overview of the status quo of television heritage (Table 1) – the status quo as well as the perspectives of accessing the television heritage under the influence of the new archival varieties on the Web.

	Discovery	Viewing	Reproduction	Use (rights clearance)
Change of access through user-trends on the web	Collaborative knowledge generation, databases and catalogues that are maintained by users or the industry. Fragmented, but searchable online collections.	Easy, although not always free of charge. Commercial and user-operated services to stream or download content.	Not necessary because permanently retrievable. Once released on the web, digital contents are easy to store and copy.	Contents can be watched individually as well as collectively, e.g. presented in a lecture. Rights clearance is still challenging if the originator is vague.
Recommendations for archival institutions	Expansion of network-based (overarching) clearing houses for archival databases (e.g. Moving Image Collections, German network of media libraries).	Creation of remote access possibilities to digital collections for educational or non-profit purposes, with viewing opportunities in libraries and educational facilities.	Willingness to compromise between the TV business, science and cultural institutions to improve the availability of archived television on a wide base.	Improving legal certainty in the use for research, education and non-profit cultural work.

Table 2: Significant changes of access to television heritage through the use of the Internet. Source: own survey.

The interviewed representatives of public institutions emphasize the mediating role of voluntary memory organizations which would have the task of striving for the strengthening of democratic access to the television heritage – also and especially with an international scope due to the prevailing restrictions on the part of the television industry. The large gaps that still exist despite the active user engagement and the increase of available programming material by the industry can be reduced by access provision to public archives, libraries and museums in a curated, and therefore structured, manner. At the same time, the cultural leading function of heritage institutions in managing access to the television heritage is paradoxically threatened by their engagement on the Internet as one of too many access providers. They have to contend with the difficulty of procuring their previously exclusive position within the information infrastructure of society. Because of the virtual ubiquity of information and audio-visual content, archival collections can no longer be seen as a unique feature:

> Perhaps the digital context will reduce the intensity historians invest in the archive and perhaps they will less likely fall under the illusion, characteristic of earlier generations, that archives contain the truth of the past (Poster, 2008, p. 21).

The Internet might intensify the effect of a possible rejection of professional conservation bodies by the users, which has grown out of the attractive ease of access to a fractional area of heritage:

> The problem right now is that people really want the information on the fingertips on the Internet. Having to come to a facility physically is a barrier. The proliferation of something like *YouTube* shows that people are posting many things that were hard to find or see before with that regards to copyright. That's the way the young generation likes to do research (Mark Quigley, cited in Kramp, 2011, p. 303).

All of the investigated public organizations are considering the extension of their services into the virtual online sphere and are defending their leading function in the discourse about television history in order to expand their outreach. Bruce DuMont, from the Museum of Broadcast Communications, expects this to the beginning of a new era of enlightenment which would offer a unique opportunity for the television heritage management to reach the social classes that are educationally disadvantaged and to interest those who might otherwise not visit a cultural institution through the attraction potential of televisual pictures and sounds and the Internet:

> I think we may be going into a greater period of enlightenment and most importantly in sharing information and willingly wanting to participate (DuMont, cited in Kramp, 2011, p. 304).

The growing use of audiovisual content on the Internet suggests that the opportunities and risks of access for the established institutions lie close together. While the current prohibitive and cumbersome access alternatives on sites (in archives, libraries and museums) are challenged by the net-based sharing principle, the television heritage is no longer understood solely in its original programming form in the care of certified preservation institutions, but as a decontextualised clip material instead, and the increase of interest for the television heritage and the overwhelming willingness of the users to participate come to mind. The interviewed representatives of the public memory institutions are therefore optimistic that they can sit at the forefront of the movement and ensure that their legitimacy will continue.

Some institutions have already taken an active step in that direction: the Paley Center for Media has a slowly-growing number of excerpts – lasting three to five minutes – in a virtual preview archive in order to allow its global user-base to watch parts of the collection that could formerly only be accessed behind the museum's walls. Those first tentative attempts are complemented by broader initiatives such as those of the "Internet Archive" project (www.archive.org) which, for instance, provides access to the complete special news programmes of ABC, CBS, CNN, Fox, NBC and the BBC on September 11 and 13 2001 in its special collection "September 11 Television Archive". Even if such singular projects still play an exceptional role, some memory institutions like the Museum of Broadcast Communications are trying to build a consistent online platform for registered users who are granted access to an already considerable online inventory of digitised programmes:

> [W]e are certainly in the forefront to make that happen, given our aggressive use of the Internet: We were the first institution to take our holdings to catalog them on the computer. We were the first to offer our card-catalog completely online, search ability online. We have over 300 hours of streaming media on our website. So we feel that we have really pushed the envelope in making content accessible to as wide an audience as possible. [...] We are a museum that is not just about tapes, but what's on the tapes. When you think about the bigness of that idea with the Internet, with streaming media, we can be providing content that has nothing to do with the history of radio and television, but it has everything to do with the history of divorce or Civil Rights (Bruce DuMont, cited in Kramp, 2011, p. 306).

Therefore, some "lighthouse institutions" in heritage management confidently profile themselves as pioneers in advancing their Internet activities with the retrieval of audiovisual material from anywhere. In addition, the paradigms of access within the organizational structures of the public administrators of the televisual corpus are inevitably changing: the institutions are transforming in a figurative sense into virtual "self-service hold shelves" (Balas, 2006, p. 39) from which the user can select provided content as he pleases. The search and retrieval of electronic records are no longer dependent on the services of qualified personnel, but are possible day and night and independent from the opening times of an archive, a library or a museum. The risk of relativising itself as a physical place of remembrance through its own website cannot be denied, nor can the potential to attract new target groups through the online channel and interest them in a visit to the physical facility. Therefore, the benefits and disadvantages of the virtual opening balance each other.

It is without question that the Internet encompasses a vast variety of remembrance work, knowledge resources and activism regarding television culture. However, what is (still) missing is an overview that ensures their development opportunities and guides them in constant orbits. From this perspective, public memory institutions need not be afraid of marginalisation, but they can be attributed to a stronger position than ever before because the lively, but unstructured and fragmented, television discourse on the Web requires coordination and orientation services, which effectively can only emanated from independent institutions with applicable expertise. This makes a commitment of public heritage institutions not only possible but absolutely necessary in order to create interfaces and clearing houses cooperatively, and in order to support existing participatory information services in creating a reliable infrastructure of access to knowledge and the audiovisual heritage.

Conclusions

Silverstone's understanding of the "mediapolis" was used as a foundation for the discussion of the changing configurations of accessing TV's heritage at the beginning of this chapter. It can be stated that through the participation of media users, the mediatised society, as the sum of the individual connections to the past (that is to a great degree connected to television history), can shape its identity and can agree on a shared (desired) image of the future, whereas the (television) images of the past, with their large number and diverse content, form the basis for this multivocal agreement on a common future-proof consensus of communal life (Silverstone, 2007, pp. 97-98). According to Silverstone, this process is more about perspectives and therefore less about the creation of binding interpretations of the past. However, as a consequence of archival restrictions, the corporate television heritage management still inhibits the functionalisation of historical television content instead of encouraging it.

The user community has long sought to cut corners and present alternative solutions from which to share beloved or interesting television programming via online platforms that provide access to audiovisual content as well as to contextual information from the history of television. This online movement, which shows anarchist tendencies in the handling of fragments of TV heritage, has led to an increase in the circulation of historical TV content, even though the availability of archival material online represents only a small proportion of that housed in the depots of the television industry. Moreover, the user-driven circulation of recordings or copies is fated to discontinuity because of the counter-measures of the rights-owners (i.e. copyright and neighbouring rights). The principle of sharing, which is the significant attribute of the young archive culture on the web, is not a new quality *per se*, but rather an integral part of traditional archives management, whose function has always been to not only preserve assets, but also to permit access to them. The awkwardness of access, which results from the imperatives of corporate television heritage administration, certainly entailed the allegedly criminal aberrations among Internet users who look for alternative ways to retrieve and share historical content and therefore infringe the copyright. Innovative marketing strategies that integrate rather than fight user activism and thus achieve acceptance among the user base could help here. The Internet is a rare opportunity to dispel existing stereotypes about the archives, as described by the Australian moving image archivist, Ray Edmondson:

> With its popular connotations of dust, cobwebs and decay, of material forgotten, locked away and remote from ordinary access, the word is often a public relations liability. The perception of material, discovered in, or 'dredged up from' the archives carries no suggestion of the precision, client orientation and dynamism of a well run archive (Edmondson, 1998, p. 7).

With the ease of distribution allowed by digital technologies and an online infrastructure already in place, broadcasting archives have the ability – if supported by corporate management – to undergo an image change and benefit from the openness, interest, and passion of the activists. Only when rights owners with their rich body of archives, memory organisations (like archival bodies, libraries and museums) with their public education imperatives, as well as the general public/user base with their disparate motivations come to a model of coexistence that is not marked by economic skirmishes but on a consensus based on complementary interests can the Internet achieve its potential as an information access way into TV's past. The current situation represents, at best, a "digital preview archive" of television history – a keyhole through which only a tiny beam of light reaches into the depths of the televisual corpus.

References

Albarran A. B. 2010. The Media Economy. New York: Routledge.

Andreano K. 2007. The Missing Link. Content Indexing, User-Created Metadata, and Improving Scholarly Access to Moving Image Archives. The Moving Image, 7 (2), pp. 82-98.

Anon. 2006. Clip Culture. Economist, 379 (8475), p. 68.

Balas J. L. 2006. The Social Ties That Bind. Computers in Libraries, 26 (2), pp. 39-41.

Bjarkman K. 2004. To Have and to Hold. The Video Collector's Relationship with an Ethereal Medium. Television & New Media, 5 (3), pp. 217-246.

Blouin Jr. F. X., Rosenberg W. G. (eds.). 2007. Archives, Documentation, and Institutions of Social Memory: Essays from the Sawyer Seminar. Ann Arbor: University of Michigan Press.

Boynton R. 2004. The Tyranny of Copyright? New York Times Magazine, 25 January, p. 40.

Burgess J., Green J. 2009. YouTube. Online Video and Participatory Culture. Cambridge: Polity.

Costello V., Moore B. 2007. Cultural Outlaws: An Examination of Audience Activity and Online Television Fandom. Television and New Media, 8 (2), pp. 124-143.

Derrida J. 1996. Archive Fever. A Freudian Impression. Chicago et al.: University of Chicago Press.

Dowman M., Tablan, V. Cunningham H., Popov B. 2005. Web-Assisted Annotation, Semantic Indexing and Search of Television and Radio News. In: Association of Computing Machinery. Proceedings of the 14th International World Wide Web Conference, WWW2005. New York: ACM Press, pp. 225-234.
Edmondson R. 1998. A Philosophy of Audiovisual Archiving. Paris: UNESCO. URL: http://unesdoc.unesco.org/images/0011/001131/113127Eo.pdf (accessed 14 May 2011).
Einav G., Carey J. 2009. Is TV Dead? Consumer Behavior in the Digital TV Environment and beyond. In: D. Gerbarg (ed.). Television Goes Digital. New York: Springer Science + Business Media, pp. 115-130.
Gracy K. F. 2007. Moving Image Preservation and Cultural Capital. Library Trends, 56 (1), pp. 183-197.
Hilderbrand L. 2007. YouTube: Where Cultural Memory and Copyright Converge. Film Quarterly, 61 (1), pp. 48-57.
Howell C. 2006. Dealing with the Archive Records. In: D. G. Godfrey (ed.). Methods of Historical Analysis in Electronic Media. London et al.: Routledge, pp. 305-348.
Jenkins H. 2006. Convergence Culture. Where Old and New Media Collide. New York: New York University Press.
Kirk J. 2011. Police Shut Down Movie-streaming Portal. PC World, Online 9 June. URL: http://www.pcworld.com/article/229865/police_shut_down_ger man_moviestreaming_portal.html (accessed 4 July 2011).
Kramp L. 2011. Gedächtnismaschine Fernsehen. Band 2: Probleme und Potenziale der Fernseherbe-Verwaltung in Deutschland und Nordamerika (Translation: Memory Machine Television. Volume 2: Problems and Potentials of Television Heritage Management in Germany and North America). Berlin: Akademie Verlag.
Lenard T. M., May R. J. 2006. Net Neutrality or Net Neutering. Should Broadband Internet Services Be Regulated? New York: Springer.
Lowry T. 2007. The YouTube Police. Big Media is Spending Millions on Monitors. BusinessWeek, 21 May, p. 42.
Miège B. 1989. The Capitalization of Cultural Production. New York et al.: International General.
Poster M. 2008. History in the Digital Domain. In: F. Gross, W. Marotzki & U. Sander (eds.). Internet – Bildung – Gemeinschaft (Translation: Internet – Education – Community). Wiesbaden: VS, pp. 15-30.
Prelinger R. 2009. The Appearance of Archives. In: P. Snickars & P. Vonderau (eds.). The YouTube Reader. Stockholm: National Archives of Sweden, pp. 268-274.
Rosen J. 2006. The People Formerly Known as the Audience. Pressthink (online), 27 June. URL: http://archive.pressthink.org/2006/06/27/ppl_frmr.html (accessed 4 July 2011).

Rubin N. 2007. Everything Old Can be New Again. Current.org, URL: http://www.current.org/dtv/dtv0708preservation.shtml (accessed 14 May 20 11).

Sarno D. 2007. A Dust-up Over Old TV Tapes. Los Angeles Times, 10 June, E1.

Schmalz, G. 2009. No Economy. Wie der Gratiswahn das Internet zerstört (Translation: No Economy. How the Gratis Mania Destroys the Internet). Frankfurt/Main: Eichborn.

Schulze W. 2000. Wieviel Überlieferung braucht die Geschichte? Überlegungen zur Ordnung des Bewahrens (Translation: How much Heritage Does the History Need? Reflections on the Order of Preservation). In: A. Metzing (ed.). Digitale Archive – ein neues Paradigma? Marburg: Archivschule Marburg, pp. 15-35.

Silverstone R. 2007. Media and Morality. On the Rise of the Mediapolis. Cambridge/Malden: Polity Press.

Skinnell R. 2010. Circuitry in Motion: Rhetoric(al) Moves in YouTube's Archive. Enculturation (online), Issue 8. URL: http://enculturation.gmu.edu/circuitry-in-motion (accessed 14 May 2011).

Trant J. 2009. Studying Social Tagging and Folksonomy: A Review and Framework. Journal of Digital Information (online), 10 (1). URL: http://arizona.open repository.com/arizona/bitstream/10150/105375/1/trant-studyingFolksonomy .pdf (accessed 14 May 2011).

Ubois J. 2005. Finding Murphy Brown: How Accessible are Historic Television Broadcasts? Journal of Digital Information (online), 7 (2). URL: http://journals. tdl.org/jodi/article/download/172/155 (accessed 14 May 2011).

Vaidhyanathan S. 2004. The State of Copyright Activism. First Monday (online), 9 (4-5). URL: http://firstmonday.org/htbin/cgiwrap/bin/ojs/index.php/fm/article /view/1133/1053 (accessed 4 July 2011).

Van Dijck J. 2007. Television 2.0: YouTube and the Emergence of Homecasting. Media in Transition International Conference (online). URL: http://web. mit.edu/comm-forum/mit5/papers/vanDijck_Television2.0.article.MiT5.pdf (accessed 14 May 2011).

Walker P. 2011. US Issues Piracy Warning to UK Website Owners. London: The Guardian, 4 July, p. 6.

Whyte W. S. 2001. Enabling eBusiness: Integrating Technologies, Architectures, and Applications. Chichester: Wiley.

Zielinski S. 1999. Audiovisions: Cinema and Television as Entr'actes in History. Amsterdam: Amsterdam University Press.

Fausto Colombo & Andrea Cuman

The (Old) New Value of Digital TV as a Cultural Product[1]

Introduction

The process of digitisation occurring within the media system is perceived and studied by many scholars not as harmonious but rather as a strong discontinuity within this scenario (Grasso and Scaglioni, 2010; Scaglioni and Sfardini, 2008; Wasko, 2005). We propose to take a step back, and look at how these transformations, and media convergence more generally, are shaping media to resemble more traditional forms of cultural circulation. In particular, we will try to demonstrate this by looking at the cultural value of the television product and at the ways it circulates in the digitised media scenario. In the first part we will discuss the implications of dealing with a medium such as TV from a culturological point of view. In the second we will focus on the cultural value of the digitised television product.

Any attempt to observe today's media scenario or, as Appadurai puts it, today's mediascape (1996) must face the consequences brought by the digital revolution. The diffusion of broadband technologies and the Internet, which is now "entering its mature phase" (Thomas et al., 2011), changed not only the media system as a whole, but also our relations with and through media. Media scholars such as Jenkins have advanced the idea of "convergence culture", one of the most notable attempts to grasp these processes

> where old and new media collide, where grassroots and corporate media intersect, where the power of the media producer and the power of the consumer interact (Jenkins, 2006, p. 2).

Benkler (2006), though starting from a more institutional and economic point of view, reaches similar conclusions, speaking of a "hybrid media ecology". Also audience studies scholars find themselves dealing with "people formerly known as audiences" (Rosen, 1999), and media historians have lately felt the necessity to reflect upon a "convergence media history" that attacks the assumption that people can be solely a film or television *user*:

---

1 This chapter discusses on reflections started in Colombo (2009, 2010) and evidence drawn from various qualitative researches on television conducted by OssCom (Research Centre on Media and Communication, http://centridiricerca.unicatt.it/osscom).

> While likely print, movies, radio, television, and new media should never have been thought of as separate histories, the insistence of context now forces [...] to note relations among and between the various sites of information and entertainment (Staiger and Hake, 2009, p. ix).

In other words, most media scholars have to deal with the increasing complexity brought about by the digitisation of media.[2]

If we ask ourselves "what are media today?", we will probably find many different answers, none of which completely fits today's constantly and rapidly evolving scenario. The same challenge is faced when we look at specific media, such as television, radio or cinema, and ask ourselves "what are they today?" Talking about media, and trying to understand them with conceptual tools that become dated as soon as they are shared, presents us with an increasingly difficult challenge. Therefore, we must think of television inside a wider frame, which is able to account for the deep connections that have changed or developed between the circulation of culture in the age of television and the other forms that this circulation has assumed. We must recognise and show new and old connections, recovering continuities and highlighting differences and fractures.[3] Moreover, it is necessary to be aware that the "age of TV" belongs not only to the medium itself, but to society more generally and the social changes that have more than once "bent" this medium for its needs, influencing the shifts and partial metamorphoses. This does not mean we should stop our questioning, but rather that we must ask other questions, that shift our perspective. For example: how does culture circulate in the age of digitised TV? Has this circulation any specificity? From this point of view, the main issue then becomes another: can we finally compare media to other anthropological forms of cultural circulation? Can we find in this comparison what really matters? The point then is not so much how television has evolved since its birth, but how the cultural form of television is moulded by today's technologically based cultural circulation, and therefore how television's contents circulate and assume value in this new context.

The television we knew in the last century can be regarded as a peculiar form it assumed for some decades, though forms of long-term resistance can be found. These forms regard "resistant" behaviours both of producers (for example, in the scheduled flow) and of consumers (for example, in their consumption habits), but also of the conceptual counterparts constituted by television scholars around the world. An example of this point of view, which we could call "internal", is given by Scaglioni and Sfardini (2008), who, even though able to highlight many changes that have occurred within the television system, still look within the me-

2 Amongst the most interesting, critical and cautious attempts to deal with the new media landscape is Dahlgren's (2009) in "Media and Political Engagement: Citizens, Communication and Democracy".

3 On this aspect see the recent reflections of Pasquali, Scifo and Vittadini (2010) who speak about "crossmedia cultures".

dium, keeping it as a pivotal point from which the more general consequences of digitisation cannot be grasped.[4] That is to say that, while looking at television from the internal point of view of broadcasters and "institutional" content production can still offer interesting insights on this medium, the most is happening outside, on and through it.[5]

Television as a Cultural Product

Amongst the insights that arose from within the sociology of culture, and with the awareness of the many definitions of culture proposed by scholarly tradition (Kroeber and Kluckholn, 1952), Griswold's approach offers important tools to understand the forms of cultural circulation and how meaning is embodied in cultural objects (Griswold, 1994). The two ideal types of culture she draws are considered, on the one hand, too wide and, on the other, too restrictive: too wide when referring to "that complex whole" which comprises the opinions, values and beliefs of human beings, too restrictive when considering it as "the best that has been thought and known," thus referring to art, literature, etc. She then prefers to define culture as the "expressive side of human life" and cultural objects as "shared significance embodied in form" (Griswold, 1994, pp. 11-12). But if Griswold's definition of culture is helpful to define the context from which we move our reflections, the heuristic tool she offers[6] is not able to fully fit the cultural and media industry scenario, as the cultural diamond

> reduces the industrialisation of culture to a productive specificity, whilst the technological nature of media deeply works also on distribution, consumption and socialisation (Colombo, 2003, p. 90).

In other words, if we try to look at television from this perspective we find ourselves in front of a myriad of subjects, in which the status of producers (some of whom start as technology producers or service providers to become content creators) and receivers (who themselves become content creators)[7] is challenged. We

4 See also Moran (2010).

5 For example, Spigel and Olsson (2004) or Uricchio (2004), who try to broaden the reflection by introducing concepts such as technologies, interface cultures and flows.

6 The cultural diamond is the heuristic tool proposed by Griswold to understand cultural circulation and cultural objects. Its shape resembles a baseball field, and is made up of four points and six connections. The four points are: cultural object, producers, receivers and the social world. Each of these points is connected to the others, and Griswold tries to account for the relations that occur between the four points.

7 Among the first to illustrate the new emerging forms of participatory culture is Jenkins (1992), who is followed by wider and more recent reflections in the audience studies, such as by Abercrombie and Longhusrt's (1998).

should draw many cultural diamonds, intersect them in various points and redesign them to fit the specific context brought by digitisation. Another way to see the problem of cultural circulation then is to isolate the necessary and sufficient elements that enable any given type of cultural circulation in a society. Each of these elements have been tackled by different scholars dealing with media: the studies of domestication processes (Silverstone and Hirsch, 1992) looked at the objectuality, at the interface and the incorporation of a medium, whilst ethnographical research (Lull, 1990; Casetti, 1995) has focused on the rituals and on contents as products with cultural value. The necessary and sufficient elements then, in our hypotheses, are materials, rituals and contents: materials are the physical basis through which culture circulates, rituals are the space-time co-ordinates which are able to define cultural contents in a given context, and contents are the meaningful substance through which culture is given as such.[8]

Materials: Digital Perceptions and Transforming Interfaces

When speaking of materials we mean the matter that is shaped and becomes an artefact, the physical substance that is used as medium and vehicle of the cultural flow.[9] On the productive side it expresses itself as a technique or technology, on the receptive as a mode of perception of matter.

The metaphors used to describe the television screen find their roots in cinema (when speaking of "small screen" as opposed to the big cinematic screen) or in the "window on the world" metaphor.[10] But thinking about the relation between the manipulation of matter and perception, we feel another example fits better: the stained glass windows in churches. Because there is no doubt that cinema and television have in common artificial light, whereas only cinema projects a light on the screen, pushing it away from the viewer. In television, however, light comes to the viewer: it is a revelation, not a projection. Pixels, like stained glass illuminated by light, bring their signs to the viewer's eye. This perception is the recognition of a signal, not an evidence of reality (this is why the window metaphor only partly

8 A similar attempt to redefine the emergent digital media landscape is made by Casetti (2009), who recognizes that speaking about "media" is no longer enough. Therefore, he tries to start a new reflection on media studies by conceptualising practices, environments and discourses as fundamental elements of the digitally mediated communication and putting them in relation with the "symbolic stances" of human beings.

9 It is not exactly the material dimension of a ready-to-use object and its affordances, as in Silverstone's (1994) double articulation, but rather the material nature of technology itself, the matter which is made of, and which varies between different techniques, as some of the following examples will try to illustrate.

10 An interesting account on the window metaphor throughout history is given by Anne Friedberg in "The Virtual Window. From Alberti to Microsoft" (2006).

explains itself). The arrangement of matter and its articulation is always a signal, an indicator for the viewer of a possible passage to another dimension, distinct from everyday life (but only rituals will enable this dimension through the streaming of cultural flows). The nature of matter and its configuration is not unconcerned (maybe it is for this reason that TV, with its matter made of pixels, is so close to those thresholds that anthropology recognises as entrances to other worlds or universes).

Moreover, digital technologies have brought to us many more "revealed" signs that we experience on many different screens (be it looking at a TV show on the computer or surfing the Web on the television screen). Television interfaces, for example, have been deeply influenced by digital technologies.[11] According to Stefanelli (2010), the surface of the screen then becomes both an "impalpable threshold" and an "autonomous planar space" where designers, graphic designers and artists have exercised their skills to create a multi-layer aesthetic which borrows much from computer aesthetics.

> But the cultural complexity of the surface also reveals the anthropological meaning of the television interface, for example in the differences between East and West.[12]

Rituals: Experiencing the New Mediascape

The second essential element for cultural circulation are rituals. Because if, from one point of view, in human societies any object can be culturalised, from the other, culture is not everywhere and always, but is given in certain places and times. According to the sociology of culture, and Griswold in particular, even a piece of bread can be a cultural object, but to be so, it must be recognised as such. To convey or consume a cultural content one must find oneself in a recognisable ritual (or we could easily mistake an ancient megalith for a huge rock, or a painting for a simple figurative element). Rituals are space-time environments through which culture flows; they are bottlenecks in which contents are regimented. Culture is not always given, but any communication could be defined as such if provided with codified rituals.[13] If, since its beginning, television has offered a universal culture that could be accessed with no literacy, thus undermining the cultural value of television, ethnographical research (Lull, 1990; Casetti, 1995) has observed many ritualised forms of consumption which worked in the opposite direction,

---

11 In what could be defined as a constant "remediation" process, as pointed out by Bolter and Gruisin (1999).

12 Stefanelli (2010, pp. 41-44) well describes the dialectic of the "material" dimension of the screen as interface, and the dialectic between its transparency and opacity.

13 On the ritual dimension of media see, for example, Dayan and Katz (1992), or Couldry (2003).

assigning television a specific cultural value. There is no culture without materials or contents, or without rituals, because cultural transmission is not everywhere and always, but only in socially assigned places and times. To investigate the ritual dimension then becomes an essential ingredient for whoever wants to understand the deep nature of the medium: no longer a precise technology or cultural form, but a mediation area between subjects, in socially given times and places.

Television has accelerated a deep transformation of social rituals in advanced societies: the bottlenecks through which contents flow and are distributed change the bottleneck's very substance. We can ritualise objects and make them culturally meaningful, but the opposite process is also possible. We can de-ritualise cultural objects, making them lose their cultural value. This de-ritualisation can occur in two ways: disregarding the cultural nature of a product (this traditionally occurs through cultural devaluation as in racism), or by introducing cultural classifications (for example, between high and low culture). In our case, television has undergone a process of value reduction that in different periods has assumed different forms – a conflicting process in which the perception of television as a cultural object was either asserted or denied. In monopoly conditions, television recognises itself as low culture but tries to resemble high culture, mainly through the type of contents broadcast (for example, theatre performances), but also by means of a strong ritualisation of consumption through very rigid programme scheduling. Commercial television, however, (being more similar to American radio or TV in its early days), immediately declines to be high culture and prefers to be entertaining and popular, using popular ritualisation forms in its schedules; in each case, the obstacle which made us unable to assign any value to its products was their "ephemeral" character. On many occasions, the ordinariness of the commercial television schedule has been read as an abdication from rituality. Today we are finally aware that it was the opposite: an intentional aim to make every day a festivity, a ritualisation of everyday life that for the first time made cultural flows available everywhere and at any time. We can once more see the "old" form of television as a transitional period, a land to go through to arrive at the Internet, where the "artificiality" of the TV schedule vanishes. The user no longer has to be reached, because he can reach the cultural content he wishes, manipulating an interface that is similar to the TV screen, and, though it is not the same, taking much inspiration from it. Digitisation transforms once more the ritual frame of broadcasting and consumption: programmes are available for a longer time, enabling users to choose the moment of the day, the device and the cost of consumption. As we will see in the final part of this chapter, this transformation is crucial to understand how television tends to create a cultural circuit that differs from the previous ones, which were based on the technological and organisational features of the television industry.

TV as Low Culture or no Culture

We have seen how TV as a cultural product can be read through a new perspective, but what about television products? It is not doubted that their cultural form has changed with digitisation, but in which ways? Has the cultural value of the digitised TV program changed? Once more, the "external" perspective we pointed out at the beginning becomes helpful. The new context we are facing sets new rules and new value-making processes, both from the broadcaster's point of view and for the consumer, but above all in the new forms of cultural circulation. To tackle this issue we will begin from an issue that, in the current media debate, is far from being solved and is strictly tied to the cultural form of television. It is about the difference between high and low culture, between what in the knowledge circuit is coded as "engaged", "important", "profound" and what is considered just as "passing the time", "entertaining", "light".[14] The main element to qualify the difference between what is "high" and "low" in culture is the legitimacy of the social discourse. Examples of what is considered high and low culture are therefore influenced by historical factors: what was once considered popular is now found in museums all over the world (one of the places in which culture as such finds its most evident legitimisation), and vice versa.

But television has been always considered as "low" culture. Only sometimes, in brief phases, has it been the vehicle for real intellectuals (in Italy, for example, during the years of the monopoly). The critics who are most prone to consider writing as literature, deal with this subject straightforwardly by addressing themes and languages through wider and more strategic issues. But the problem is still there, especially if we shift from TV in general to a single programme, the most similar thing in television to other types of cultural objects. The art market, for example, has clearly showed the role of supply and demand in the aesthetical perception of a piece: an object can easily become a work of art simply by socialising its economic value (Ruccio et al., 1996; Veltius, 2005). The problem is that in the circulation of a cultural object, the apparently extrinsic element that is price, strengthens the circuit, the trust in authors and buyers, the public visibility of a piece, and so on. So a cultural object is defined as such by elements regarding not only its "culturality" but also its "objectuality". Italy and its cultural market offer significant examples of the redefinition and explosion of products that were earlier excluded from public relevance: commercial TV is a good example, and on this we will say a few words.

Italian commercial TV started developing during the late seventies, and it was immediately accused of being a sort of irrelevant mash by those who supported the traditional monopolistic TV. But commercial TV knew it was good for its au-

14 On the relations between the various forms of "culture" and cultural circuits see also Crane (1992).

dience, with its content flow, advertisements, colours, variety of programmes and a new imagination. The price paid by commercial television, in the relationship between "popular" product and culture, was relevant. The product was defined by certain new characteristics, which confirmed its distance from high culture and assigned it to a sort of separate circuit from traditional cultural institutions.

The first of these characteristics is "gratuitousness", something similar to what in the eighties was defined as the "right to consume": having communicative content at no cost would become precisely a right. In this significant development, two apparently contradictory impulses converge: on the one hand is egalitarianism, the heritage of the utopian years of the late sixties, on the other is the properly consumeristic strategy which makes you dream of products just by making them apparently available.

The second point is the ephemeral nature of commercial TV. It was obvious from its technological birth that the volatility of the scheduled flow made a programme an impermanent entity (with the exception of the mysterious and often unreliable broadcaster's archives). Recording on videotape – widely available since the eighties – was a domestic practice reserved for certain types of content such as film and programmes for children. One consequence was that a critique practice such as in literature or cinema was impossible; it could only assume the form of comment, rather than be a guide for consumers (as consumption, by the time of reading, had already occurred). So, commercial TV redefined "low / popular" objects through a special relationship with the audience and, in doing so, created a new socially legitimised cultural circuit, though not integrated into the traditional one (the "high" one of school, literary criticism, recognised intellectuals and cultural institutions), and rather opposed to it. Television programmes were then legitimised without having unhinged, but rather confirming, the contradiction between high and low, between decent and unworthy. This game of opposition/legitimacy is clearly manifest in the fact that the product of this new circuit has no history or memory. It is unworthy of repetition or re-vision, unless in the form of mania, cult consumption or fandom.

A Paradigm Shift: Networking Contents and New Values

With the digitisation of media, new roles of cultural products were defined, all of them, finally without any distinction.

Look at the transformation of television today: we are beyond the (probably long-term) resistance of the traditional content flow,[15] organised around the

15 See, for example, Williams' fundamental reflections on the TV content flow (1990).

schedule with its timing and its "warranty period",[16] with its funding from advertising and the still dominant idea of the audience as the product and the program as the "hook" for the audience. Digitisation, in this sense, had at least two consequences for the television system: the first is that it changed once more the cultural form of television content through technological achievements. The second is that it gave birth to another circuit, external but related to the first, in which television contents circulate, are legitimised and valued in unprecedented ways.

Regarding the transformation of the "old" cultural circuit of free and ephemeral commercial TV, we can see that on-demand products and consumption practices on the one hand overcome the myth of gratuitousness, reconsidering the legitimacy of a price to pay and, on the other, articulate menus that offer a wide range of cultural products (movies, series, cartoons, etc.) that stand there at the disposal of future choices. The television product can be enjoyed online, perhaps through free download or streaming, in its original versions, then seen on pay-per-view or pay TV, and then, again for free, on broadcast TV, and then perhaps be bought on DVD. Regarding this emerging cultural form of television content, we want to highlight some aspects that arose from several researches we conducted recently.[17] These aspects regard the relation between the television offer and new consumption forms in a multi-device scenario, in particular as concerns user-centred time management and the emergence of a new spatial economy which are able to assign new values and new forms to the television content. This is to say that, if the "ephemeral" flow of commercial television kept the content distant from the viewer, digital television brings it nearer and makes it usable as never before.

Being able to decide when to watch a TV programme is not only a technological achievement of digitisation; it is above all a possibility to value television content through different consumption practices, in which the value assigned to the time spent watching TV defines the value of the content that will be viewed. It is the difference between time-investing and time-spending consumption practices, where the first tends to privilege contents that have a "higher value with respect to the scheduling of the broadcaster" (Colombo & Vittadini, 2010a, p. 5), whilst the second tends to become a low value consumption, to spend time that would otherwise be perceived as wasted. In this new context, where the ephemeral content

16 Translated from Italian, "warranty periods" are those periods during the year when broadcasters ensure advertisers a minimum audience share, so as to maximize profits for both (for advertisers who can count on a "mass" audience, for broadcasters who can sell spaces at higher prices).

17 In the last two years, OssCom developed a research programme on television appropriation practices by young Italians in the cross-platform scenario. The research was based on multiple methodological techniques (ethnography, in-depth interviews, etc.) and perspectives (production analysis, secondary data analysis, etc.). For more details see Colombo and Vittadini (2011a, 2011b). These researches build on wider reflections started in Colombo and Vittadini (2006) on the digitisation process of television and the diffusion of digital terrestrial technologies, a comparative study that involved different European countries.

of traditional broadcasting and the repeatable content of on-demand and pay-per-view platforms live together, consumption practices are able to define high and low value TV products. Highly valued products are more likely to be viewed cross-platform, maybe first on TV and then on DVD, computers and then stored in various ways (either a hard drive or a DVD), whilst low value products will remain in the ephemeral dimension of one-time consumption of the broadcasters' schedule.

But the new cultural form of television is also defined by re-appropriation practices which re-define the spatial economy of TV consumption. The TV sets that we once used have profoundly changed: higher quality screens and the attached digital devices have created a multi-platform environment that enables us not only to watch TV, but also to play video-games, surf on the web, etc.:

> This kind of high quality 'TV set' is then characterised by scarcity in the spatial economy of the households and different kinds of negotiations arise around it (Colombo and Vittadini, 2011a, p. 7).

In other words, what happens is that the value of a television product is also based on the choice of the space where it is to be consumed: modern living rooms and enhanced TV sets lead to a negotiated definition of the value of the TV product, while often bedrooms are devoted to more personalised forms of consumption, which mix high and low value TV products. Moreover, digitisation has brought new spaces for consumption, both outdoors and mobile, in which low investment consumption practices[18] drag the contents out of households and bedroom activities and into the public and socialised space.

> This new spatial economy then emerges as characterised by an opposition between fixed and mobile; individual and social spaces where some cross-border practices take place. While linear consumption of broadcasting (except for live events consumed through portable devices) remains a fixed practice, non-linear consumption practices of television programs used per se are the ones that cross the borders of spaces (Colombo and Vittadini, 2011a, p. 8).

In a cross-media and networked scenario, television also must then be regarded as a networked medium: on the one hand it becomes part of a wider network of devices, on the other hand TV products circulate in different forms through the Internet. In this sense, television tends to remain a compass for consumption practices, and the first place where linear forms of consumption

> seem to conserve the capacity to supply a wide range of programs shared among users and to activate other practices of acquisition or enhancement (Colombo and Vittadini, 2011a, p. 9).

18 We mean by this term those forms of consumption where viewers assign scarce economic, cultural and informative value to these time-spending practices (Vittadini, 2010, pp. 187-213).

But it is through computer and other digital devices that TV products are becoming more "tangible": before digitisation, TV products could only be consumed in realtime or by videorecording. Today we can digitally record them and watch repeatedly, download them to our devices, edit, republish and share them on different technological platforms so as to extract this product from the TV flow it belonged to. New forms of non-linear consumption, of "horizontal" circulation and manipulation of content are enabled. We are not speaking of user-generated content, which constitutes one of the acclaimed phenomena of digital culture, but of the possibility of traditional TV content entering this new digital circuit and undergoing a process of "substantiation" and value enhancement. What seems to happen is that a new bottom-up or horizontal circuit is formed,

> a kind of flow that involves the members of a social network built in real or virtual life. [...] Television programmes acquire, then, a specific social value: the more they can be *spent* socially among the peer network, the higher is their value. The more they have an identity value for the group, or they can become cult products or, on the contrary, the more they are unique and new for the group, the more they can be spent (ibid.).

This exchange value was obviously impossible to obtain in the circuit of "ephemeral" television. With digitisation it not only becomes possible, but changes the relationship between audiences and broadcasters. It is no longer the case, as when television was "ephemeral", that the greatest threat to broadcasters in the exploitation of the (economic) value of content is other television networks (commercial, local, etc.) in the most typical form of market competition. Viewers did not have the technical possibility of extracting any value from a television programme, and the social spendability of programmes was often as rapid as the airing of the program itself. With the multiplication of platforms, the abundance of supply and the permanency of content, consumers and their devices have also entered the competition, on the one hand compelling the big networks to protect their content from new forms of user consumption (in some cases defined as piracy) and, on the other, trying to extract value through new strategies (for example, through different forms of content distribution).[19] The ephemeral flow of content is lost, in favour of a permanency that is given by the archiving of content (by broadcasters and

19 An interesting case in Italy is the TV show *Lost* created by Lieber, Abrams and Lindelof. It was initially aired in prime time on RaiDue, but due to low audiences was re-scheduled late at night. But the series had in the meantime gathered a number of fans who started viewing the show on the Web and, thanks to the *quasi*-realtime subtitling of the series by the Italian fansubber communities, they were able to watch the new episodes shortly after they were aired in the US. This compelled content providers and broadcasters to shorten the time between the broadcasting of the show in the US and in Italy in an attempt to exploit an audience that had moved away from the traditional television "platform" to new forms of consumption, where the consumption and exchange value of the users did not coincide with the economic one of broadcasters.

their on-demand menus, by users and their personal archives), and by the constant circulation of this content in the Internet. The content can certainly be nourished by these archives, but finds its always-on condition in what is known today as the Internet "cloud", a digital space that can be regarded both as ephemeral (in its possibility of access, in its technical nature) and absolutely permanent: websites, social networks and peer-to-peer networks that are all part of this "cloud" and keep content alive and circulating thanks to the users' activities. This leads to elongation of the life-cycle of the cultural product, giving it a new economic value which is based on a price, which in turn makes some issues essential (time priority, exclusivity, the ability to remain on the market, the capacity to be socially spendable, etc.). It is a silent revolution where TV products apparently change only slightly, where productive processes are relatively slow and therefore less noticeable.[20] But it is nonetheless a revolution, which has eventually reshaped the entire television system, leaving it at the centre of a publishing model in which the public taste, its temporal and cultural needs are worth more than before. It is a return to traditional forms of cultural markets, from art to cinema and publishing.

This transformation is perhaps the most radical among the many that we are observing, even if it is deeper and therefore less noticeable. Of course, even at the level of the product and how it is "staged", this entails serious consequences: the menu screen of an on-demand TV looks more like a library shelf than a television screen of the eighties (as, in a similar way, online newspapers are becoming more similar to television and its flow of content and "liveness"[21]). But it is not an epochal turning point, only the closing of a bracket which lasted for a few decades: a very short time from a historical point of view. However, in so doing, brackets open to something new, because a television product that looks more like a book in its offer and symbolic consumption recalls to mind a cultural dimension that erases "otherness".

The availability of broadband and a wide range of contents that have become more similar to traditional cultural objects, require a re-evaluation of the latter in the universal library of knowledge. Rather, it should be noted that it is the products of high culture that have lost their references and guarantees, the traditional academies with the failed intellectuals who sanctioned its supremacy. So, in the sixties a television product had a cultural value only if it related to high culture, quoting it or staging it; in the eighties and nineties its value was based only on its

---

20 In fact, the ways broadcasters use the capabilities of digital technologies are innovative more in the different contact strategies with the audience than in the assimilation of the new value of the television product (Beyer et al., 2007).

21 Especially during events, where "live channels" of audiovisual streams are opened. But a curious case, from this point of view, is thaxt of the Italian daily newspaper "La Repubblica" and its new website claim: "In diretta con l'Italia", which could be translated as "Live from Italy", a claim that years ago we could only imagine in television channels and certainly does not belong to the publishing industry.

success; today all products seem to have the same value, with the problem that there are no shared references. What identifies the cultural value of television products today is no longer the distinction between high and low, but the lack of institutions to determine the difference. We just recognise that value is there, and that quality is possible somewhere waiting for someone to tell us where.

References

Abercrombie N., Longhurst B. 1998. Audiences: A Sociological Theory, Performance and Imagination. London: Sage.

Appadurai A. 1996. Modernity at Large: Cultural Dimensions of Globalization. Minneapolis: University of Minnesota Press.

Benkler Y. 2006. The Wealth of Networks: How Social Production Transforms Markets and Freedom. New Haven and London: Yale University Press.

Beyer Y., Enli G., Maasø A., Ytreberg E. 2007. Small talk makes a big difference: Recent Developments in Interactive, SMS-based Television. Television and New Media, 8 (3): pp. 213-234.

Bolter J. D., Gruisin R. 1999. Remediation: Understanding New Media. MIT Press.

Casetti F. 1995. L'ospite fisso. Televisione e mass media nelle famiglie italiane. Milano: San Paolo.

Casetti F. 2009. I media dopo l'ultimo big bang. In: Che Fare? La TV dopo la crisi. Cologno Monzese: Link/RTI.

Colombo F. 2003. Introduzione allo studio dei media. Roma: Carocci.

Colombo F. 2009. Parentesi Chiusa. La scoperta del valore culturale dei prodotti TV. Mash-up Television. Cologno Monzese: Link/RTI.

Colombo F. 2010. L'eredità culturale della televisione. In: F. Colombo (ed.). Tracce. Atlante warburghiano della televisione. Cologno Monzese: Link Ricerca/RTI.

Colombo F., Vittadini N. (eds.). 2006. Digitising TV – Theoretical Issues and Comparative Studies across Europe. Milano: Vita e Pensiero.

Colombo F., Vittadini N. 2011a. Reconfiguring TV: The Cross-Platform Scenario. Working paper.

Colombo F., Vittadini N. 2011b. Digital Television: Appropriation Practices and New Cultural Forms. Working paper.

Couldry N. 2003. Media Rituals. London: Sage.

Crane D. 1992. The Production of Culture: Media and the Urban Arts. London: Sage.

Dahlgren P. 2009. Media and Political Engagement: Citizens, Communication and Democracy. Cambridge: Harvard University Press.

Dayan D., Katz E. 1992. Media Events: The Live Broadcasting of History. Cambridge: Harvard University Press.
Friedberg A. 2006. The Virtual Vindow. From Alberti to Microsoft. Cambridge: MIT Press.
Grasso A., Scaglioni M. 2010. Televisione Convergente. La TV oltre il piccolo schermo. Cologno Monzese: Link Ricerca/RTI.
Griswold W. 1994. Cultures and Societies in a Changing World. Thousand Oaks: Pine Forge Press.
Jenkins H. 1992. Textual Poachers: Television Fans and Participatory Culture. London: Routledge.
Jenkins H. 2006. Convergence Culture: Where Old and New Media Collide. New York: New York University Press.
Kroeber A., Kluckholn C. 1952. Culture. New York: Meridian Books.
Lull J. 1990. Inside Family Viewing: Ethnographic Research on Television's Audiences. London et al.: Routledge.
Meyrowitz J. 1985. No Sense of Place: The Impact of Electronic Media on Social Behaviour. New York: Oxford University Press.
Moran A. 2010. Configurations of New Television Landscape. In: J. Wasko (ed.). Companion to Television. Oxford: Wiley-Blackwell.
Pasquali F., Scifo B., Vittadini N. 2010. Crossmedia Cultures: giovani e pratiche di consumo digitali. Milano: Vita e Pensiero.
Rosen J. 1999. What are Journalists for? New Haven: Yale University Press.
Ruccio D., Graham J., Amariglio J. 1996. "The Good, the Bad, and the Different": Reflections on Economic and Aesthetic Value. In: A. Klamer (ed.). The Value of Culture: On the Relationship between Economics and Arts. Amsterdam: Amsterdam University Press.
Scaglioni M., Sfardini A. 2008. Multi TV. L'esperienza televisiva nell'età della convergenza. Roma: Carocci.
Sennett R. 1977. The Fall of Public Man. New York: Knopf.
Silverstone R. 1994. Television and everyday life. London: Routledge.
Silverstone R., Hirsch E. (eds.). 1992. Consuming Technologies: Media and Information in Domestic spaces. London: Routledge.
Spigel L., Olsson J. (eds.). 2004. Television after TV: Essays on a Medium in Transition. Durham et al.: Duke University Press.
Staiger J., Hake S. 2009. Convergence Media History. New York et al.: Routledge.
Stefanelli M. 2010. Questioni di superficie. In: F. Colombo (ed.). Tracce. Atlante warburghiano della televisione. Cologno Monzese: Link Ricerca/RTI.
Thomas F., Vittadini N., Gòmez-Fernàndez P. 2011. Cultural Influences on the Adoption of Web 2.0 Services. In L. Haddon (ed.). The Contemporary Internet. Frankfurt am Main: Peter Lang.

Uricchio W. 2004. Television's Next Generation: Technology/Interface Culture/Flow. In L. Spigel & J. Olsson (eds.) Television after TV. Essays on a Medium in Transition. Durham et al.: Duke University Press.

Veltius O. 2005. Talking Prices: Symbolic Meaning of Prices on the Market for Contemporary Art. Princeton: Princeton University Press.

Vittadini N. 2010. La televisione (digitale) nel contesto crossplatform. In: F. Pasquali, B. Scifo & N. Cittadini (eds.). Crossmedia cultures: giovani e pratiche di consumo digitali. Milano: Vita e Pensiero, pp. 187-213.

Wasko J. 2005. A companion to television. Oxford: Blackwell Publishing.

Williams R. 1990. Television. Technology and Cultural Form. London: Routledge.

Eleonora Benecchi & Giuseppe Richeri

TV to Talk about. Engaging with American TV Series through the Internet

Introducing Fandom

In January 2008, Kristin Dos Santos of *E! Online*, posted a public survey asking to vote for the best TV series of the current season, the winner would have become the protagonist of a special section that was about to be opened on the website. The list provided by the reporter included the most popular series on the major networks. "Supernatural" won with 29% of the votes, followed by "One Tree Hill" with only 28%, both series are on CW, the younger and minor network (Dana, 2008) among those included in the survey, while other networks series were far away from these percentages.[1] Kristin decided to delete these two titles from the survey and opened it once again suggesting that the first results were not reliable and that two fandoms so active and interactive must have been "influenced" or "encouraged" by the network. This statement generated a series of public and private responses to the journalist that suggested, more or less aggressively, the same message: never underestimate the fandom.

This episode proves to be interesting for two reasons. On one hand it underlines how even a self proclaimed "TV fan" (in her official bio on *E! Online* website she included among her after-school activities "channel surfing and TV-guiding") who built her fame by chatting with other TV fans and dishing out scoops on their favourites TV shows, can underestimate the online community of fans. On the other hand, it suggests that the formation of an active and organized community of fans online does not depend on the amount of attention that the network addresses to the audience, in terms of promotional investment and advertising, but from the type of activities in which it involves this community. This is indeed one of the main hypothesis of this article and the analysis of fan practices proposed here will in fact try to demonstrate and substantiate this idea.

With respect to the first point, the "fan" has always been an elusive category and in addition a very low number of research studies have attempted to answer the question: "what does it mean to be a fan of a television series?" As noted by Le Guern (2002) many of the studies regarding this phenomenon show in one instance the tendency to overestimate their practices, what the fans are doing, and

1 Supernatural Legend Community, Fandom Campaigns (Materinal derived from Personal Communications Interviews in May-June 2010).

secondly to underestimate the social determinants, who the fans really are. Even more difficult is to study television fans meeting on the Internet and building communities: studies often remain at the theory level and can rarely document how fans discuss and take possession of the television programmes they are devoted to, or the ways they interact with TV producers. A relevant exception is represented by Jenkins and his research on fandom (1992, 2006 and 2007), though as stated by Jensen and Pauly (1997) to speak about the audience as an interpretive community, as Jenkins is doing, means to focus on media texts and on texts generated by fans in response to them, while neglecting the social processes through which these meta-texts are processed.

The ethnographic research on the audience failed to explain the culture or the sub-culture created by the communities born online around a television series (Nightingale, 1996). Even though it could be remarked that this was due to the early and preliminary stage of online communities in the mid-90s, when the first ethnographic studies were realized, it must also be taken into account that these analyses often lack an extensive and participant observation and more often this ethnographic approach results in short visits to the online communities and individual interviews or focus groups with some of their members (Moores, 1993). In actual fact, the Internet seems to have simply enhanced features usually attributed to the traditional fandom: creativity, critical approach, participation (Fiske, 1989; Gripsrud, 1998; Jenkins, 1992). From another point of view, the Net seems to have helped change forever the nature of TV series' fandom: some researchers even argue that the online fandom has evolved from a niche phenomenon into a mass subculture (Wu Ming 1 and 2, 2006).[2] Specifically, processes of lauding, preserving, collecting, scrutinizing and being passionate about a popular TV series are becoming both increasingly prevalent and increasingly culturally acceptable (Carey, 2005). Moreover, increasing numbers of viewers are being empowered and encouraged to become deeply passionate about TV series, or in other words to become "fans" (Caldwell, 2008; Hills, 2003; Baym, 2000). Instead of replacing television, the Web seems to have embraced it (Naughton, 2006) and with regard to TV series in particular, the Internet seems to have turned Television into a shared event for the most passionate viewers, who have become plugged-in, both sharing the show with the other viewers and becoming invested in the creation of the show itself (Jenkins, 2006; Hills, 2003; Baym, 2000).

Therefore, the online fandom related to television products, cannot be analysed only as a simple online community, whose organizing principle is the interaction through a particular network, or as a "television" community organized around a particular text, but must be analysed as a community managed through a series of structured and routine practices (Hanks, 1996; Lave and Wagner; 1991). In such

2 Wu Ming is a group of anonymous authors who call themselves Wu Ming and put a number after this name in order to distinguish the different papers.

perspective, what can be defined as online fandom, that must not be confused and overlap with the fan community in general, is here studied through the analysis of communication patterns and thus as a "community of practice" (Hanks, 1996).

Shifting the focus on the production side, the second point highlighted by Kristin Dos Santos' "case", the American television series are increasingly built to encourage an active participation of a passionate audience, which show off more and more insistently on display through spectacular forms of exploitation and consumption. Take for example the way in which Fox has promoted and distributed "Glee", one of its top series, not only building an appointment with the weekly episode for the average viewer, but also addressing the fan viewer, disseminating the text with "Easter eggs"[3] and inviting him/her to deepen the viewing experience by participating in side experiences and by purchasing collateral products. This type of strategy has transformed even the "traditional" fiction into a rich multimedia and multiplatform experience, based on the convergence between the TV set and the computer screen through the Net and specifically social networks. The high involvement required in this case, as well as in others, had the goal to draw attention to the experience of being members of an audience, by enhancing contents and creating groups of "believers" (hence the name of "loyal communities") emotionally tied to a television product and therefore willing to invest, even economically, in franchises' products related for genre/trend to their object of desire (Benecchi and Colapinto, 2011).

Fandom practices generated spontaneously coexist with new fan practices that are both planned and required by the production and the promotion strategies put in place by the television industry (Caldwell, 2008). According to Johnson (2006) and Carey (2005) the very success of the participatory model proposed by recent American television series, helps to emphasize the presence of an increasingly active and alert audience that is less and less willing to be satisfied with the classical and mono-medial narratives.

Watching a television series today means to see it several times on different media platforms, (TV, streaming, downloading, DVDs), to discuss narrative developments and deepen ones knowledge on the Internet, to collect related material. Activities once reserved just for the fan audience are becoming more and more widespread and at the same time increasingly culturally acceptable. Viewers are given the opportunity to become "experts" of one or more television series, turn into a crowd of "fans" and above all "collectors" (Wu Ming 1, 2006). It must also be remarked that "reception processes" alone, such as watching the program on various platforms or collecting collateral materials, are not representative of the whole spectrum of fan involvement with TV series, since other participatory

---

3 Easter Eggs are special contents traditionally hidden inside DVD (today, more generally speaking, they are positioned between audiovisual texts) and they are not indicated among the technical characteristics of the titles. They are often short movies, gags or backstage that can be found among the various menues building each DVD.

mechanisms have emerged as fans and producers work together, through social networking sites, promoting and "spreading" a specific TV series contents and materials. Internet provides, indeed, tools that can alter the whole experience of watching television, giving access to the process, not just the results, and talking about American TV series specifically, many scholars have pinpointed the crucial role of the Internet, in reshaping the relationship between Television Networks and their audiences (Askwith, 2007; Jenkins, 2007; Caldwell, 2008; Roscoe, 2004). The choice to focus on these participatory mechanisms also implies a vision of fandom as a social experience where dedicated fans are often driven to connect with other fans (Baym, 2000; Le Guern, 2002).

This brief overview on the investigation field, makes it clear that the choice to focus the attention on the online fandom of American television series is based on two assumptions: firstly, American television industry appears to be one of the most inclined to use the Internet in order to create an engaging experience for customers and to foster the interaction with viewers by building on existing communities such as Facebook, Twitter and dedicated forums (Askwith, 2007). It's indeed evident that recent TV hits have been created, promoted and distributed not only taking into account the existence of an active and Internet-based fan audience but also relying on this same audience in order to make the series an international success. With special regard to these programs, patterns of interaction connecting producers and consumers are both representative and innovative, that is why they represent such a particularly interesting object of study; secondly, many American TV series have already succeeded in generating deep, perpetual audience engagement (Askwith, 2007; Caldwell, 2008; Jenkins, 2006; Hills, 2003; Baym, 2000) and in relation to this type of product it has been possible to note the emergence of a new kind of passionate audience, no longer a niche as in the past, but widespread and mainstream, so that the figure of the fan could be seen in these cases as the prototype of the daily behaviour of the audience in the near future.

My interest in investigating television fans activities outside the original country of production, is supported by evidence collected during an initial phase of desk research which underlined how: American TV series are built in order to be "distributed" to an international audience (Caldwell, 2008) and transmedia storytelling (Jenkins, 2004) is frequently presented as the best way to engage a wide and dispersed audience in the new media landscape, even though the effectiveness of transmedia storytelling outside the original country of production is still to be demonstrated through specific studies; many "post-network" TV series have been distributed with success in European countries, giving birth to strong fandom communities (Scaglioni, 2006; Porter and Lavery, 2006), even though the role of those fandom communities in promoting the TV series and in making their transmedia storytelling accessible to a wider and less engaged audience is still to be explored.

Building from this, the analysis proposed here focuses on online "promotional activities" pursued by fans of American TV series based outside the original country of production and specifically in Italy. The choice to focus on online "promotional fan activities" for TV series is based on the hypothesis that it is by keeping a show in the popular discourse, by creating spreadable media that can be shared among fans and non-fans alike, that fans can have a grassroots effect on media (Booth, 2010). In this study, I planned to analyse online fan texts and productions aimed at influencing the production and distribution of my cases of analysis with a specific focus on productions considered as "promotional materials".

The aim of this work is to build an initial framework of analysis that can be the guide for further comparative research. From a methodological point of view, as remarked by recent studies of the field (Booth, 2010; Jenkins, 2009), empirical data about fans are usually drawn from one of two sources: a) ethnographic studies of fan communities, looking at groups of fans (fandom) in order to see how the interaction between fans helps to stimulate interest in the objects of study; b) analysis of fan-created texts, looking at the creations of individual fans to form inductive conclusions about fandom.

With regard to the first type of analysis, many are the reflections on the need to change research methods and the researcher's role in the new context of online fandom studies (Pasquali et al., 2010; Livingstone, 2007). With regard to this last point, the crucial question remains an old one: if a researcher in online fandom must be a pure observer (Andò and Marinelli, 2010) or must in fact sustain the double role of researcher and user (Vellar, 2010; Beer and Burrows, 2007).

With regard to the second type of analysis, User Generated Content are acknowledged as an ideal source to research the new "connected audiences" (Boyd, 2008), but it's still unclear how to process all the materials available online.

I choose to use here a mixed method approach. I've started my analysis with a desk research focusing specifically on fan promotional activities that require a constant use of the Web in order to be completed and aim at influencing the producers of the show. The desk research was performed using many different sources: scientific papers, online resources both official and unofficial (such as TV critics blogs and websites, website for American Networks and TV series, articles and online reviews). This phase of analysis helped me substantiate with the collected data that fans try to influence TV series production, promotion and distribution in two main ways: a) actively petitioning a Network to keep a show on air; b) participating online in official and unofficial promotional activities in order to keep communication about the show alive.

The desk research was followed by an explorative analysis of the fan activities previously identified focusing on the materials produced by members of two online fan communities based in Italy (Supernatural Legend and Subsfactory) and the organization of two different fan campaigns supporting American TV series and involving Italian fans in their activities (specifically Save Jericho and Supernatural will not get lost). In order to better shape and design the on-going taxonomy about online promotional activities pursued by fans of American TV series, I've performed an empirical analysis using traditional ethnographic tools such as biographic interviews (specifically, between 2007 and 2010, I've conducted 30 face to face interviews with fans of different age groups, located in different regions and members of different fandom) and observation (while attending three editions of "Telefilm Festival", a national convention dedicated to TV series, I've collected speeches and thoughts produced by the fans themselves).

The methodology chosen for this study respond to the need of giving back to the researcher's subjectivity a central role in the research process, profiting from its ability to observe and establish relationships (Vergani, 2011).

In the first part of this article, I will provide an overview of the online fandom's activities identified through the analysis of the online materials collected during the desk research and explorative phase. This preliminary overview of fan practices will be based on the comparison with the traditional categories of fan productivity identified by Fiske, developed by Hobson (1982) and subsequently revised by Jenkins (2007) and Baym (2000).

Thoughts collected during the observation phase and comments made by fans during the interviews are used to support and confirm my findings. This is useful to better understand what it takes to be part of an online fandom and specifically what types of practices a fan usually performs to participate in the community gathered around a specific TV series.

In a second part of the article I will focus specifically on collaborative fan works produced through intense interaction and group collaboration in order to illuminate a shift in the manner of textual creation introduced by online fandom. Again, references to the collected materials and fan quotations will be used to corroborate my analysis.

In the conclusions I will provide a catalogue of practices, involving fans of TV series, based on their most relevant traits.

What Does it Take to Be a Fan Nowadays

According to the traditional model of fandom studies, fandom is characterized by two major tasks: the discrimination in the selection of the object to "worship", and various forms of productivity, ranging from the acquisition of a style and a look proposed by a certain object of worship, up to the active contribution to a social production of contents related to the same object of consumption.

The first element emerging from the analysis proposed here is the confirmation that online fandom is made up of "expert consumers", able to trace "generic" histories and interpret the product they are passionate about, putting it in relation to countless preceding examples and previous products inside the same genre. This aspect shows on one hand a strong sense of inter-textuality that characterizes the fan viewer and on the other hand the connected dissolution of boundaries around the text itself. The Italian community of fans of "Supernatural", for instance, dedicated an entire section of the forum hosted by "Supernatural Legend Website" to summarize and explain both the quotations and excerpts from American urban legends and horror stories explicitly mentioned inside the story and the links between each episode and a series of texts, books, comics, songs or films, not explicitly mentioned but still recognized as connected with the series itself. A following analysis of a popular international fan site, "Television Without Pity", made it clear that this is a practice shared by the international fandom of the series: inside the forum devoted to Supernatural after each aired episode, a discussion about quotations and citations from pop culture products is opened. Moreover the TV series is designed to encourage this type of inter-textual fruition, as proven by the fact that its creators are constantly "playing" with their audience including traces of other texts, more or less explicitly, inside each episodes.

This kind of inter-textual reading developed by fans, and often labelled under the umbrella term of "nitpicking", challenges the idea that the meaning and the sense of a TV series are built in and pre-existing: in the case of "Supernatural", specifically, it is only by reading the original text in connection with other texts that fans can find the "real" meaning and value of the series, becoming at the same time an interpretive community.

Regarding the characteristic act of "discrimination", it is true that during the recording of the interviews fans were able to point out explicitly which series and even which characters in particular they were fans of; it is also true that not always the enthusiasm for a TV series becomes all-encompassing, defining the lifestyle and identity of an individual. In other words, there are many ways to be devoted that do not necessarily result in fandom, and since they are more frequent and widespread they are perhaps the most interesting to study. From this point of view, interviewed people stressed the plurality and variety of their consumption: nobody stated to be passionate about only one series, more often I could observe what could be called a voracious consumption that lead respondents to follow an average of twelve series per week and to become members of at least three different fandom at the same time.

Talking about the "productive" aspect identified as the main feature of the traditional fandom, we can say that the Internet has witnessed a progressive development of this aspect so that this has become the main characteristic of modern fan clubs surrounding a successful TV series (Couldry, 2004; Cova et al., 2007; Boyd, 2008; Booth, 2010). From the interviews and the observation carried out on the field, it emerged that it is not enough to show a comprehensive knowledge on the series in question to be identified as a "true" fan, but it reveals to be more and more important today to show an active and productive capacity in relation to the series itself.

In the Italian fandom of American TV series, for instance, I was able to identify several communities of "fansubbers", or fans that created Italian subtitles for TV series and distributed them to other fans. One of the most active and structured community is connected to the website "Subsfactory.it" created in 2006. This website is different from other "archives of subtitles" because of the public purpose stated by its founder to "create a community around fansubbers and users of subtitles". From the beginning and not surprisingly, the website required a subscription, even if a free one, unlike other stores of Italian subtitles where the presence of lurkers, users who visit the site in a hidden way, without declaring their presence, is quite common according to the fansubbers interviewed. The registration tool was in fact introduced because the website was supposed to become

> an active meeting place not a passive one, where you only enter to download a subtitle and turn away immediately, a place in which people interact, try to coordinate with one another in order to avoid double work, where new people can become fansubbers, collaboration between different groups of fansubbers is enhanced, and, perhaps, a place to meet new people with the same passions and make new friends (Superbiagi, personal interview, 2009).

This is an activity which, as confirmed by other fans involved in the project and interviewed during the last national Telefilm Festival (2010), requires a strong and continuous commitment (4 to 5 working hours over night on average for each TV series translated), a set of linguistic skills and high level techniques, and a constant relationship with the other members of the group. The majority of those interviewed confirmed having studied at a language university course or to work as a translator, right alongside the efforts of fansubber. Those without any background in the field reveal a rather strong passion towards the specific product they have to translate or a passion for TV series in general. What is interesting to note, however, is the fact that the fansubber does not necessarily translate the series he is a fan of, although a good level of knowledge of the series is still needed for it to be translated. The fansubber is therefore both a provider of collateral material on the series and a user of the material provided by other fansubbers. Connected to this issue is the tendency to experience the activity of fansubbing as a socializing collective experience: on one hand, a fansubber rarely translates an entire episode of a series, usually the original text is split between several fansubbers, coordinated by a supervisor in charge of standardizing the various translations into a single final text, on the other hand the fansubbers interviewed claimed that with the passing of time the link with the group has become more important and decisive than their passion for TV series themselves. In this case, the additional work produced for the object you're a fan of, is done in the perspective of providing a "service", either to the TV series itself or to other fans.

The phenomenon of fansubbing also illuminates a specific trend of online fandom: watching the TV series through the web. It is a practice confirmed by all respondents, even if it is implemented with varying degrees of intensity: ranging from streaming viewing or legal downloading via the official websites, to the illegal downloading for personal use, distribution of illegally downloaded material through a protected forum, up to the collective viewing, via platforms such as MSN and Skype, of illegally downloaded episodes.

It is a fact that was also confirmed by examining public documentation on the historical strike of the WGA[4] writers called for the 5 of November 2007 against the producers of the AMPTP[5]. Beneath the highlighted need for a new production model able to face the emergency situation and correct inefficiencies, a shift in the distribution of TV series towards more immaterial channels (streaming and downloading in particular) is also signalled so as the need to confront with a situation in which television viewing decreases in favour of new consumption patterns.

In recent years, in fact, illegal trading of TV series through programs that use the Torrent[6] protocol is represented by different sources (Brennan, 2008; Bauder, 2007; Ellis, 2007; Harmetz, 2000) as a growing phenomenon and as a clear indication to move the distribution of these products to more immaterial channels and mobile services.[7] In 2007 nearly 50% of people who used the Torrent protocol have downloaded a TV series at least once, according to a survey made by Van der Sar (2007). Considering that in the reference year it was estimated that only 10% of the content available on the network was formed by television series, the weight of this share is even more evident. In 2008 another survey reveals that 40% of the downloaded torrent files correspond to episodes of television series (Van der Sar, 2008). This is quite interesting considering that the most illegally downloaded TV series in 2010 were offered free on legally specialized portals

---

4 The Writers Guild of America (WGA) is the Trust of the writers of the show (cinema, radio, and television).

5 The Alliance of Motion Pictures and Television Producers (AMPTP) is the organization representing the interests of almost four hundreds films and television producers which have a contract with WGA's workers.

6 Through the Torrent protocol, one of the peer-to-peer protocols, the files sharing is not central and users can download files one from the other. The protocol consents download and upload at the same time: even who has got a partial file can send it to the others while he is downloading missing parts.

7 According to the estimates of Torrentfreek, the higher numbers of 2007, linked only to the Web search for torrent Minova are: "Heroes" 2,4 million, "G"alactic" 706,000, "Lost" 705,000, "Dexter" 435,000. One of the series like "Lost", though, can reach up to 10 millions of download considering the joint esteems of different torrent sites.

such as Hulu[8] or on the official TV Networks websites (Van der Sar, 2010). With special regard to this last result, the author of the report stresses that the majority of illegal downloads of series available for free on official channels seems to be made outside the United States, which is also confirmed in a study by De Kosnik (2010). However all fans interviewed living outside the United States, notice that another aspect has to be emphasized: the problem is not so much in the delay of the original material distribution, but the fact that fans feel they deserve an archive easily accessible, of high quality and long-lasting, all characteristics ascribed to pirated files and not to the official ones.

Fan Productivity and Collaborative Fan Works

Being a member of the audience today means no longer necessarily receiving a message by a producer, since the mutual mingling of the roles between producer and consumer requires the acquisition by members of the audience of a varied set of skills that leads them to create performances by themselves (Fiske, 1989, p. 147; Le Guern, 2002, p. 179). In this context, the fan is no longer generally described as an active spectator, but as the most active among the spectators of a serial product (Gripsrud, 1998, p. 113). What emerged from the analysis of the materials produced by fans in relation to American television series during the sample period considered, however, was not so much a hyper-connected consumer activism but rather a tendency to team productivity. The case of materials produced and activities undertaken by the fandom of "Jericho" (CBS) once the TV series cancellation was announced it is a shining example of this trend. The practice of sending individual letters or signing petitions to convince a production not to end a TV series or to resume the production of an already cancelled one, is nothing new in the context of television fandoms and it is not an activity specifically related to the online fandom. There's no doubt that the Internet has increased the collaborative nature of this practice and has helped to emphasize the social aspect of fandom. Specifically, during May 2007, CBS announced the cancellation of "Jericho" giving as a main reason the disappointing ratings, they were around 9.5 million viewers per week. In the same month the "Save Jericho Campaign" started, involving fans of the series from different parts of the world, grouped or not into local communities, and culminated in sending 9 million pounds of nuts to CBS as a homage to the last words uttered by Jake Green, leading character of the series, who was in turn quoting the historic statement, "Nuts!" by General Anthony McAuliffe during the Battle of Bastogne. Fans were playing

8 Created in 2007 through a joint-venture between NBC Universal and New Corp giving the availability of NBC and Fox programs, but always cable TV series. In respect to others, Hulu is a free platform supported by advertising.

with the ambiguity of the term used in English both to indicate the peanuts and to blame an act or a request considered "crazy". The sending of peanuts to CBS was a way of telling the Network executives they were "crazy" cancelling the series. Campaign promoters took advantage of the free and ubiquitous nature of the Web during the organization period, but they also demonstrated to be conscious of the economic nature of the talks they wanted to involve the Network into when putting themselves "for sale" as final consumers for investors in advertisements. The highly structured nature of fandom that promoted the campaign became obvious when they signed agreements with the online retailer Nuts.com to ensure a more aggressive campaign. CBS's response, posted on the forum of the campaign on the 6th of June 2007, was the renewal of the series for seven "test" episodes. On one hand, the letter of responsibility sent by Nina Tassler, president of CBS, to the fans recognizes their passion and the structured nature of their community while on the other hand invites them to find "new audiences" for the series.

This type of campaign is not an isolated phenomenon but is becoming increasingly common, showing as fans build their practices also on the success of activities undertaken earlier by other fans or other fandoms. In January 2008, after the writers' strike, the CW announced that "Supernatural.tv", one of its "historical" series would be put on Thursday at nine o'clock, against the first episode of the new season of "Lost" and the old episodes of "CSI". Concerned about the potential decline of Supernatural's Ratings, around an average of 3 million viewers per night, since the TV series was tied to fight with two crowd pleasing and successful shows with well established audiences, the leaders of two amateur sites with a large fanbase, "Supernatural.tv" and "Winchesterbros.com", started a campaign entitled "Supernatural Will Not Get Lost" to earn "followers" to their favourite program:

> despite what producers do, (the director and Kripke), Supernatural is not promoted enough by the Network, so it's up to us fans to promote it for them. Let's talk to our friends and to everyone we know (official statement published in the website campaign).

In other words, the promotional activities of the fans in this case is compensating for a lack of action by the network. The official launch of the joint campaign was then scheduled for January 31, 2008, the date starting from which "Supernatural" is supposed to enter into a direct competition with the new season of "Lost". The description of the campaign and its goal was featured on many fansites and forums: to increase the ratings for "Supernatural" by promoting the series through online and offline word of mouth. In reality this is just an instrumental objective to the primary campaign: ensuring the renewal of the series for a fourth season. This goal, however, was never made explicitly clear during the campaign, but only in correspondence responding to its closure.

From January 31 to February 21, in the four weeks when "Supernatural" was on-air in the same time slot with "Lost", before the supposed suspension, the organizers of the campaign posted regular reminders on various sites and forums, inviting fans to watch the series and use word of mouth to convince friends, colleagues and family to do the same. Reminders had a standard format divided into modules with different functions: the first sentence was a recall followed by a section that provided information on the planning of the series. This was useful to allow the target of the campaign to complete the requested behaviour: watching the series on TV, getting family and friends to watch it themselves through the information on programming and finally send positive feedback to the Network. It also presented a section where some data was provided to emphasize the criticality of the situation and the consequent need for intervention by the fans. This was followed by an explicit expression of the targeted behaviours required. The memo concluded with the recall of all the headlines of the campaign and was "signed" by the two sites promoting as the seal and guarantee of reliability and seriousness of the initiative.

After a week from the start of the first phase of the campaign, the organizers offered fans the chance to send some fan arts that were planned to be disseminated on the net later on. At first the fan arts sent to the two referral emails were posted within special spaces created inside sections dedicated to images ("gallery") of "WinchesterBros.com" and "Supernatural.tv", but the invitation was to download them and store them on other sites and forums. The goal was to raise awareness about the campaign "Supernatural Will Not Get Lost" and at the same time to promote the series. This method was not competitive, there were no prizes for the sending fan arts, but it was a joint effort to "get the word out" in a creative way.

On February 11, ten days before the supposed suspension of the series, the organizers of the campaign aimed to reach a new target: the online journalists. The headline of the campaign was re-formulated into "Do not Fear the Reaper!" in an explicit reference to the title of the series that would replace Supernatural in the time slot of Thursday evening, "Reaper". There is also a second level of meaning to this: the Reaper of souls is a character that the two main protagonists have to face several times during the three seasons of the show, always winning this fight. This reference could only be clear to long-time fans of the series, but obviously the organizers were also interested in them following the campaign. The objective of this phase was to encourage journalists to talk about "Supernatural" in their articles online. The chosen channel was the email transmission. Again, the organizers provided a format and some rules to follow: the email should be sent to the list of journalists provided by the organizers; should contain as many references as possible to "Supernatural" in the form of questions, comments, opinions, since "the force is in the numbers"; emails should not contain insults or attacks to the TV network or its executives, also they should not denigrate "Reaper", nor threaten or attack journalists in any way. The Campaign was closed when the network announced the renewal of the series for a fourth season. On the organizers' websites official thanks to all those who participated in the campaign were posted.

The analysis of campaigns conducted through the Internet organized by fans of a certain TV series conducted through the Internet could also help to open a discussion on the effectiveness of television programming and other methods of transmission of the series proposed by the American Networks, the accuracy of Nielsen data, the "economic" value of a fan audience and its ability to have an influence on producers.

Online Fandom: a Catalogue of Practices

Fitting inside a general scheme the trends and examples described in this work we can say that the activism of the online fandom unfolds in three major areas related to the debate on the series, the relationship with its producers and the creative work on its contents:

The first area, and range, of fan's activism includes the speculation and assessment of future developments in the narrative plots, a phenomenon born with soap operas and therefore called "soap gossip", which is today extended to the fandom of TV series. This practice is not only providing the typical "oral" discussion, but in particular the form of participated and written discussion made possible through forums and mailing lists. It is also interesting to note how, thanks to the net this practice is becoming more visible and therefore fans "culture" is becoming increasingly central to the strategy of the production companies and TV Networks, thus ensuring these elements are indeed interesting to investigate.

A recent case is the purchase of one of the most important fan site dedicated to television, "Televisionwithoutpity" (TWoP), by the network Bravo (NBC Universal owned). The forum was started in 1998 by Tara Ariano (nickname Wing Chun) and Sarah Bunting (nickname Sars) to recap "Dawson's Creek", and was never meant to influence producers but rather to impress the community built around the single TV series (Andrejevic, 2008). As Baym observed in her study of soap opera fans,

> the Internet gives fans a platform on which to perform for one another, and their informal performances might please fans more than the official ones do (Baym, 2000).

Even if TWoP owners were working for free and pure pleasure at the time, it does not mean they were not producing value, as shown by the average of one million unique visitors per month and the publication of a crowd-sourcing book. The Network Bravo realized that the site helped draw viewers to particular shows and allowed them to build up social and information capital that increased their commitment to viewing and decided to add TwoP to its "portfolio of linked digital assets" and make it an "online brand destination" (press release, 2007). The stress on the economic value for the Network of TWoP was so evident that the creators of the board felt the need to assure their follower through a public announce on the front page (2007) that:

> we can snark it up as hard as we ever did, this change will only make the site bigger and meaner.

Despite the tagline "Snark it up, spoil the Networks" is maintained, TWoP experiences a complete redesign which transforms it into a typical entertainment site, or "a garishly unusable pile of synergy-laden things" as stated by one of the long-time users (Jim Connelly, personal post 10 April 2008), alienating the most passionate fans so as the original owners who decided to abandon the site after only one year since the starting of the partnership with Bravo.[9] This example is a proof of the differences between the ways corporations and fans understand the value of grassroots creativity and enlighten the fact that monetising fandom activities is not so easy since fandom is transforming the objects of commodity culture into gifts that are anti-commercial both by nature and by principle (Booth, 2010).

The second area, or range, of online fans activism concerns the activation of channels of dialogue with production companies and broadcasters through the practice of sending emails or online petitions and the promotion of online campaign in support of TV series to ensure renewal. A preliminary desk research has enlightend that this practice was already present in traditional fandom. And yet it found inside the online fandom a new location and the ideal support to emphasize its "social" aspect and make more visible the mechanisms of "pressure" available to fans. In this case the "gift" given away by fans is a surplus of labour created not for other fans but to compel the original media property to notice their indebtedness and do something for the fans in return. These gifts show the original producers that there is a lively fandom for their product, and possibly encourage the producers to make more.

Last but not least, the third area, or range, of fan activism is expressed in the creation of collateral materials connected to the TV series. We speak here about various activities, brought together under one umbrella: from the creation of short stories or even entire sagas in which the series' characters live new plots and interface with each other in entirely new ways, phenomenon known by the name of fanfiction, up to the production of videos that retrieve the original audiovisual material of the show and remount it in new ways, or phenomena such as fansubbing. Unofficial fan works and official ancillary content both contribute to the narrative world of a TV series but once again the drive behind their creation and exchange is fundamentally different since from a fan perspective

9 Bravo Press Release (2007).

value gets transformed into worth, what has a price becomes priceless, economic investment gives way to sentimental investment (Jenkins, 2009).

Within each of these areas we can distinguish between three types of register used by fans: an expressive register, following which the discussions and promotional materials are the production of a pure "labour of love", in other words they are produced to express the passion towards the TV series without any further intention if not the gratification obtained when receiving positive feedback; a social register in which the link with the television program serves more as a "heritage" that allows to enter a community and participate in a number of fulfilling side events to the vision of the program itself; and a service register, in which the discussions and materials are designed and manufactured to provide an explicit "service", for instance by promoting the series or making it more easily available or "readable" to its fandom.

One of the most interesting aspects that emerged during this study was the emphasis on the collaborative nature of the productivity of online fandom, and the tendency of the social register to prevail on the purely expressive one, even in the cases of "aggressive" practices aimed at "pressuring" the production. In other words, the case examined shows that, when talking about online fandoms, fans' involvement with the community itself becomes gradually more important than the discussion of the show or the production of fan labours.

References

Andò R., Marinelli A. 2010. Fare ricerca sul fandom on line. I fan italiani e le serie tv. In: S. Monaci & B. Scifo (eds.). Sociologia 2.0 Pratiche sociali e metodologie di ricerca sui media partecipativi. Napoli: Scripta Web, pp. 189-222.

Andrejevic M. 2008. Watching Television Without Pity. The Productivity of Online Fans. Television & New Media, Sage Publication. URL: http://tvn.sagepub.com/cgi/content/abstract/9/1/24 (accessed 18 May 2011).

Askwith I. 2007. TV 2.0: Turning Television into an Engagement Medium. Doctoral thesis, MIT, September. URL: http://cms.mit.edu/research/theses.php/IvanAskwith2007.pdf (accessed 18 June 2007).

Bauder D. 2007. More than Half of DVR Viewers Skipping Commercials, Nielsen says in first ad study. New York: Associated Press, 31 May. URL: http://www.mediapost.com/publications/?fa=Articles.showArticle&art_aid=102908 (accessed 18 June 2007).

Baym N. 2000. Tune in, Log on. Soaps, Fandom and Online Community. Sage: London.

Beer D., Burrows R. 2007. Sociology and, of and in Web 2.0: some Initial Considerations. Sociological Research Online, 12 (5). URL: http://www.socreson line.org.uk/12/5/17.html (accessed June 2007).
Benecchi E., Colapinto C. 2011. When Music Meets Television and Social Media. Idols in the Ara of Convergence and Participation. In: Z. Vukanovic & P. Faustino (eds.). Managing Media Economy, Medi Content and Technology in the Age of Digital Convergence. Lisbon: Media XXI Publishers.
Booth P. 2010. Digital fandom: New Media Studies. New York: Peter Lang Publishing.
Boyd D. 2008. Taken Out of Context: American Teen Sociality in Networked Publics. PhD dissertation, University of California Berkeley, School of Information.
Bravo Press Release. 2007. Bravo Announces First-Ever Media Acquisition with TelevisionwithoutpitY.com, the Online Destination for the Discerning TV Junkie, 13 March. URL: http://www.thefutoncritic.com/news.aspx?id=2007031 3bravo01#ixzz1OK626sFe (accessed 24 January 2010).
Brennan S. 2008. Producers Try New Creative and Financial Models. The Hollywood Reporter Online, 29 May. URL: http://www.mediaxchange.com/dramasu mmit/documents/HR.pdf (accessed January 2008).
Caldwell J. 2008. Production Culture. Durham NC: Duke University Press.
Carey J. 2005. What Good Are the Arts? London: Faber and Faber.
Couldry N. 2005. The Extended Audience. Scanning the Horizon. In: M. Gillespie (ed.). Media Audiences. The Open University; Media audiences. Maidenhead: Open University Press, pp. 184-222.
Cova B, Kozinets R., Shankar A. 2007. Consumer Tribes. Oxford: Elsevier.
Dana R. 2008. It's No Gossip, Ratings Slip Threatens CW Network. Wall Street Journal, 16 May. URL: http://online.wsj.com/public/article/SB1210895460430 97065.html (accessed 13 July 2008).
De Kosnik A. 2010. Piracy is the Future of Television. White Paper, University of Berkeley. URL: http://convergenceculture.org/research/c3-piracy_future_televi sion-full.pdf (accessed 20 March 2011).
Ellis R. 2007. Studio Source: 'I Helped Upload TV Pilot'. AllYourTV.com, 3 August.
Ferguson M., Golding P., Cultural Studies in Question. London: Sage.
Fiske J. 1989. Moments of Television: Neither the Text nor the Audience. In: E. Seiter et al. Remote Control: Television Audiences and Cultural Power. London: Routledge.
Gripsrud J. 1998. Television, Broadcasting, Flow: Key Metaphors in TV Theory. In: C. Geraghty & D. Lusted (eds.). The Television Studies Book. New York: Arnold.
Hanks W. 1996. Language and Communicative Practice. Boulder, CO: Westview Press.

Harmetz A. 2000. 'They're Rumors, not Predications', Los Angeles Times, 29 October.
Hills M. 2003. Fan Cultures. London: Routledge.
Hobson D. 1982. Crossroads: The Drama of a Soap Opera. Massachusetts: Methuen.
Jenkins H. 1992. Textual Poachers. Television, Fans and Participatory Culture. London, Routledge.
Jenkins H. 2006. Convergence Culture. Where Old and New Media Collide. New York: New York University Press.
Jenkins H. 2007. Fans Bloggers and Gamers: Media Consumer in the Digital Age. London: Routledge.
Jenkins H. 2009. If it Doesn't Spread, it's Dead (Part Three): The Gift Economy and Commodity Culture, Confessions of an Aca-Fan: The Official Weblog of Henry Jenkins, 16 February. URL: http://henryjenkins.org/2009/02/if_it_doesnt_spread_its_dead_p_2.html (accessed 18 February 2009).
Jensen J., Pauly J. 1997. Imaginig the Audience: Losses and Gains in Cultural Studies. In: M. Ferguson and P. Golding: Cultural Studies in Question. Thousand Oaks, California: Sage Publications, pp. 155-169.
Johnson S. 2006. Tutto ciò che fa male ti fa bene. Mondadori Strade Blu: Milan.
Lave J., Wenger, E. 1991. Situated Learning: legitimate peripheral participation. New York: Cambridge University Press.
Le Guern P. 2002. Les cultes médiatique. Paris: PU Rennes.
Livingstone S., 2007. The Challenge of Changing Audiences: or, What is the Audience Researcher to Do in the Age of Internet. In: R. Andò (ed.). Audience reader, saggi e riflessioni sull'esperienza di essere audience. Milano: Guerini.
Marshall P. D. 2004. New Media Cultures. London: Arnold.
Moores J. 1993. Interpreting Audiences: the Ethnography of Media Consumption. London: Sage.
Naughton J. 2006. The Age of Permanent Net Revolution. The Observer, 5 March.
Nightingale V. 1996. Studying Audiences: the Shock of the Real. London: Routledge.
Pasquali F., Scifo B., Vittadini N. 2010. Crossmedia Cultures. Giovani e pratiche di consumo digitali. Milano: Vita e Pensiero.
Porter L., Lavery D. 2006. (Cult)ivating a Lost Audience: The Participatory Fan Culture of Lost. In L. Porter & D. Lavery. Unlocking the Meaning of Lost. Naperville: Sourcebooks.
Roscoe J. 2004. Multi-Platform Event Television: Reconceptualizing our Relationship with Television. The Communication Review, 7 (4), pp. 363-369.
Scaglioni M. 2006. TV di culto. La serialità televisiva americana e il suo fandom. Milano: Vita e Pensiero.

Superbiagi 2009. Fansubbing and Online Fan Communities. Interview, Personal communication, May.
Van Der Sar E. 2007. BitTorrent in Focus: TV Series are Hot. URL: http://www.TorrentFreak.com (accessed 17 May 2007).
Van Der Sar E. 2008. 50% of All BitTorrent Download are TV shows. URL: http://www.TorrentFreak.com (accessed 14 February 2008).
Van Der Sar E. 2010. Top Most Pirated TV-shows. URL: http://www.Torrent Freak.com (accessed 17 February 2010).
Vellar A. 2010. Grown Up with Dawson (and the) Desktop. Riflessività e partecipazione per un'indagine multi-situata del fandom telefilmico. In: S. Monaci & B. Scifo (eds.). Sociologia 2.0 Pratiche sociali e metodologie di ricerca sui media partecipativi. Napoli: Scripta Web, pp. 223-240.
Vergani M. 2011. Folksonomy nel Web, tra utopia e realtà. In: S. Tosoni (ed.). Nuovi media e ricerca empirica. I percorsi metodologici degli Internet Studies, Milano: Vita e Pensiero, pp. 115-139.
Wu Ming 1, Wu Ming 2 2006. Mitologia, epica e creazione pop al tempo della Rete. L'Unità, 31 December.

Part III

The Transformation of Television: Contemporary Perspectives

Juan Miguel Aguado, Claudio Feijóo, Inmaculada J. Martínez & Marta Roel

Mobile Television, a Paradigmatic Case on the Uncertainties and Opportunities of the New Media Ecosystem

Introduction

The television ecosystem is currently undergoing a perfect storm of technological, regulatory and sociological changes. Internet and screen diversification are amongst the drivers that contribute to these transformations. Defining television according to a given technology or a given way of consumption is no longer possible. Television contents can now be accessed through different windows and in different ways. In this context, mobile television constitutes a relevant example of the problems related to the convergence between broadcasting schemes and telecommunications services.

Despite the expectations raised in the mobile and media industries, mobile television has not yet found its commercial expression in Europe, and even in the case of countries like Japan or Korea, where it has enjoyed relative success, mobile TV finds itself in an impasse (Feijóo, Gómez-Barroso and Ramos-Villaverde, 2010).

On the basis of techno-economic ecosystem analysis (Babe, 1995; Ballon, 2007), we carry out an exploratory approach to the causes for the modest adoption of mobile TV in Europe. This may prove helpful in understanding not only the specificity of the mobile environment, but also the kinds of challenges that multiscreen convergent television services are obliged to face in order to survive radical changes in their ecosystem. In fact, beyond the technical and economic barriers related to the very specificity of the mobile service, the two ecosystems involved (mobile content and television) share obstacles inherited from similar confluent sectors (audiovisual production and broadcasting, handheld manufacturers, application providers and telecommunication operators).

In view of this, we identify four types of barriers that burden mobile television development. On the supply side, the potential hurdles are in the technical (infrastructures, standards), economic (revenue models, content availability) and normative/institutional domains. On the demand side, a gap between companies' perceptions and users' expectations is highlighted: mobile actors must understand the relevant users' expectations, the circumstances in which they intend to use the service and how they integrate video consumption rituals in different contexts and through different technologies.

The current development of mobile broadband, the diffusion of more video-friendly mobile devices (Feijóo and Gómez-Barroso, 2009) and the popularisation of new distribution channels, like application and content stores, may facilitate a second chance for mobile television take-off. In these circumstances, an exploration of the nature and resistance of the barriers that have undermined the development of mobile television to date may provide relevant knowledge about the changing conditions of the mobile environment. Simultaneously, these changing conditions affect the confluence of mobile and television ecosystems, opening the possibility of transferring mobile broadband distribution models to television standards (Ballon, 2009; Braet and Ballon, 2008).

Conceptual Frame and Method

Mobile television has been widely approached from technical (Hee Shin, 2005), regulatory (Curvey and Whalley, 2008) and consumer perception (Vangenck et al., 2008) viewpoints. Soon after its technological implementation, promises about it becoming a relevant driver for the social adoption of mobile services dissolved. Understanding the contrast between a positive user perception and a poor real adoption (Schuurman et al., 2009) became a challenge that addressed analysis to the social and techno-economic perspectives.

The conceptual frame of political economy (Babe, 1995; Freeman and Soete, 1997) combined with the ecosystem approach (Iansiti and Levien, 2004) offers a theoretical ground that is able to merge the technological, social and economic aspects. In this context, the revision of concepts such as "business model", "value chains' and "value nets" is a valuable resource in the attempt to cope with the complex and rapidly changing environment of emergent technology-related economics (Bowman, 2003; Fransman, 2007; Ballon, 2007). The value chain based ecosystem approach provides a useful descriptive framework allowing the clear delimitation of actors, interaction nets, strategies, problems and challenges within the heterogeneous and mutable environment of mobile content (Uglow, 2007; Ballon, 2009). Under such changing conditions, a descriptive approach constitutes a necessary base for specific analysis as well as in designing viable future scenarios that may guide decision-making by the involved actors.

In this paper we identify the general structure of the mobile television ecosystem, outlining the core relations amongst the different actors along the value chain of mobile television content and services. A comparative study of mobile television adoption models and figures in USA, Europe and the Asia-Pacific area contributes to primarily differentiate the socio-technical conditions under which mobile television is implemented. Assuming that mobile television ecosystem faces techno-economic barriers beyond socio-cultural differences, a SWOT (Strengths and Weaknesses, Opportunities and Threats) based approach is adopted to depict the nature and reach of the present hurdles for mobile television. The arguments in this paper are part of a research project on the mobile content ecosystem supported by the Spanish Ministry of Science and Innovation (CSO2009-07108-SOCI).

Mobile Television Ecosystem

Although they are both part of a merging context, mobile television is still a different concept from mobile video (including mobile access to Internet video as part of it). Mobile television involves those services and/or technical structures that make it possible to consume conventional television content on a mobile device.[1] In this sense, mobile television is a paradigmatic case of adapted content (Feijóo, Maghiros, Abadie and Gómez-Barroso, 2009), i.e. a content that, being produced for (and according to the standards of) an existing media, is transformed in order to facilitate its consumption on a different platform.

Television content can reach the mobile handset in two ways: through broadcasting via the adaptation of a standard television signal to mobile handset reception, or through the mobile broadband connection offered by a cellular communication network. The difference here is not merely technical, since television content value chains and the role of actors involved differ significantly whether they are attached to one provision model or another. In the first case there are no conceptual differences with conventional television, except for the necessary adaptation of the signal to the mobile receiver. As in standard television, the main income depends upon advertising or, in the case of codified channels, on subscription or pay-per-view fees. The value chain in this case remains similar to that of conventional television (Figure 1).

1 A mobile device is any digital computing device conceived for easy portability and able to access advanced mobile communication networks (3G, 3.5G or beyond). This includes (but does not end in) smartphones, tablets, 3G and WiFi connected game consoles, netbooks, etc.

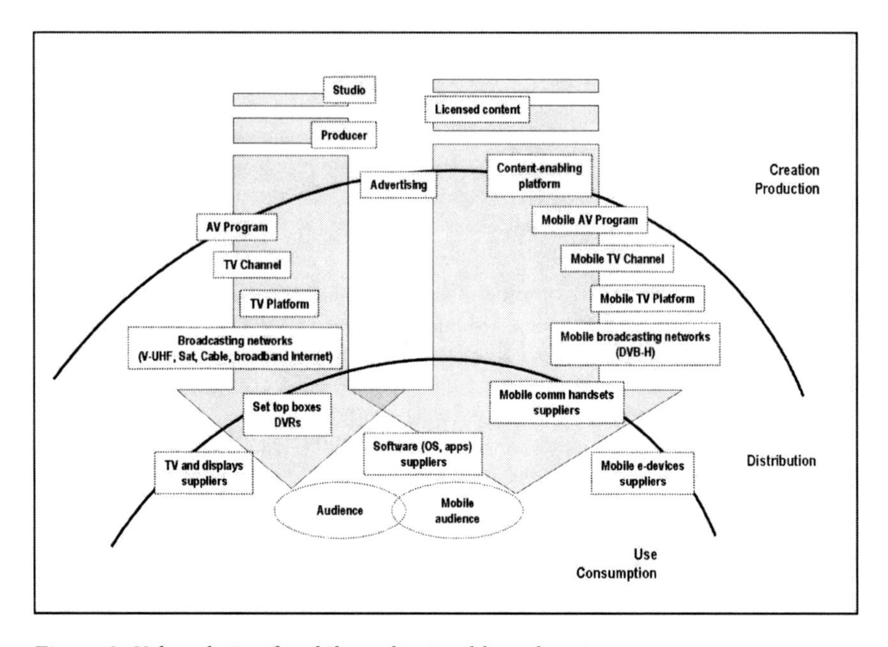

Figure 1: Value chain of mobile audiovisual broadcasting

If the service is based on a network provided by a mobile communication operator, television content may reach the audience in three different ways. The most commonly used is individual transmission to a given mobile device (unicast). The problem here is that, due to the high bandwidth requirements of video signals, a sudden confluence of connections may pose serious congestion problems to the network. This is a relevant risk, for example, in the case of live-time contents such as sport transmissions (Kretzschmar, 2006). A possible alternative to this problem is the simultaneous diffusion of video signal to a group of users (multicast). However, in this case receivers lose some of their freedom of choice about content. The third alternative is content exchange among users (peer to peer), but this option is still under research to date.[2]

From the business viewpoint, in this second case there is a confluence of two separate industries (with different interests) in close relation to the final consumer (Figure 2). Besides advertising, subscription fees and pay per view income models, this option includes also the possibility of bundling mobile television in a pack of voice and data services, similarly to the triple play models in fixed broadband.

2 There are also some proposals to integrate both technologies for an optimization of the resources required. They are, as of 2011, still undergoing testing.

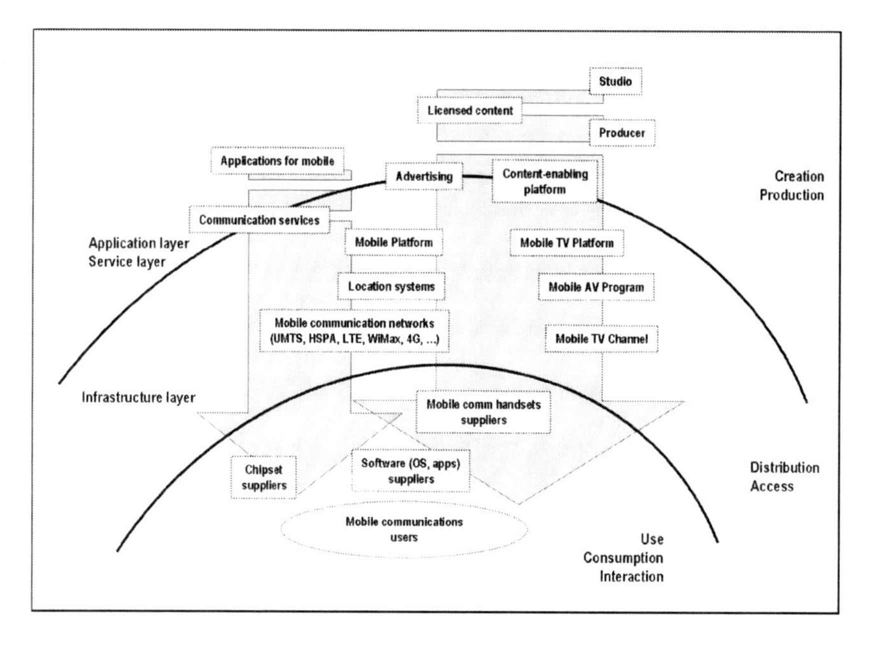

Figure 2: Value chain of mobile audiovisual communications

In both models companies struggle to place value (and thus associated income) in the part of the process they keep under control. Experienced television companies and web portals focused on information and television content prefer the first model, in which the network has a mere instrumental role. Mobile communication operators, however, try to impose their conditions on the broadcasting sector, using television as a lure to make their broadband services more attractive (Uglow, 2007).

Relations among all the actors involved are complex. Operators and device manufacturers obviously need agreements with broadcasting industries and content producers. But the other way round is also true: even in the first model mobile industries are present. In the DVB-H Forum companies involved in the European mobile TV standard depict this dependence as follows:

> There are many reasons why a co-operative approach [between broadcast sector and telecommunications operators] may be judicious. For example, many countries have mobile phone models which see the phones being subsidized by the operators, and to have mobile TV on such phones would require some co-operation between the mobile TV operator and the telco. Billing is going to be a key element to the success of mobile TV, and telecoms operators typically have sophisticated billing infrastructures in place (DVB-H Forum, 2007).

The picture gains even more complexity when considering the increasing relevance of platforms and software developers in the mobile content ecosystem (Ballon, 2009). The growing influence of platforms in determining emerging business models, like application and content stores, and their strong relationship with handset and technology providers may tip the scale of the mobile television ecosystem in favour of the mobile sector.

Mobile Television Adoption in Europe

The general picture of mobile television business model implementation shows two different, but convergent, business models. While the Asia-Pacific (mainly Japan and South Korea) area starts from a free-to-air broadcasting mobile television model, the case of Europe and the USA is that of a fee-based model structured around the mobile operator (Enter/Idate, 2010). In Europe's model, the role of mobile operators remains dominant since their practical monopoly over the end-user relationship affects the entire value chain (Prario, 2007). The strengths of this model lie in the ability to generate stable and sizeable revenues via subscriptions and to integrate mobile TV services into a pack of value-added services by mobile operators. The capacity of mobile operators to subsidize and promote compatible handsets may also encourage their subscriber base. While the challenge for the European model concerns the capacity to generate a stable, wide enough base of subscribers, the challenge for the Korean model is to generate stable revenues from a wide base of users.

Looking into figures compiled by the authors from public industry sources referring to 2009, mobile TV penetration was already much more modest than regular predictions, and furthermore it shows a scenario of strong regional fragmentation in regard to technological alternatives. For instance, in Europe there were 5 million users, most of them under the telecommunication network model, with the exception of Italy. In the USA the situation was similar, with 10 million users. Japan and Korea, however, counted some 60 and 30 million users respectively, with a clear orientation towards the broadcasting model (despite a significant increase of the supply based on the telecommunication model). The differences are conclusive: the mobile TV penetration rate of South Korea, Asia's most developed mobile TV market, was close to 20%. Yet penetration in Italy, Europe's most advanced market for mobile television services, was still less than 3%. However, despite its relative success in adoption, the cases of South Korea or Japan are still far from successful: the free-to-air model faces high infrastructural costs that cannot be covered by an unstable fragmented revenue model. For instance, in 2007, the service offered by South Korea's TDM-B standard generated revenues of 6 million USD, while its operating costs were about 40 million USD (Enter/Idate, 2010).

Looking with more detail into some of the European markets there are many more failures than success stories. In Spain, the third European country with regards to 3G penetration, there were 350,000 mobile TV subscriptions in the same year (over 300,000 in 2007) that meant an income of 19 million Euros for mobile operators (0.5% of the total television income) (CMT, 2010). Austria, Germany, the Netherlands and Switzerland all launched DVB-H broadcasting services. However, as of 2011, all of them have been closed or are scheduled to close. Significant hype also surrounded the launch of DVB-H in France – currently indefinitely postponed. In the UK, operators conducted numerous trials but have not brought the service to commercial reality. Italy, together with Finland, is the major exception in this scenario of failures. It had about 1.2 million users using DVB-H. But even in this respectable case compared to the rest of Europe, mobile television take-up has failed to meet expectations.

Barriers in the Development of Mobile Television

On the supply side, three main barriers can be identified in the development of mobile television: Technical barriers, concerning infrastructures and standards, economic barriers, concerning the definition of viable business models, and normative barriers, concerning regulation and policies in the EU. However, the more relevant barriers seem to lie in the demand sector and concern the gap between value perceptions of the service in productive agents and in users.

Technical Barriers

The first hurdle concerns the technical prerequisites for the viability of mobile television. The cost of infrastructures and the role played by the actors involved (mainly broadcasting companies and telecom operators) may make the dominant model lean in one direction or another. In Europe, as mentioned, mobile television broadcasting has not exceeded the level of occasional experiences, while mobile broadband networks are rapidly spreading and plans for a fourth generation with a substantial increase in bandwidth capacity are already being implemented (GSA, 2010).

Another technical concern refers to the standards for transmission. In the relatively immature market of mobile television, there is not yet a sufficient level of standardization in the norms of technology. On the contrary, several standards struggle to impose different norms. In the case of broadcasting mobile television, the European Commission guidelines from December 2008 call upon the industry to guarantee DVB-H-based service interoperability in every EU member nation by rolling out non-proprietary technologies. This was supposed to be the first step towards the implementation of DVB-H as the European standard for mobile television. However, this has not happened and the promised standard now remains lost in a sort of political and normative limbo (Grivet, 2009). There are, in turn, other standards, such as DVB-SH (which combines the possibilities of terrestrial and satellite diffusion), DMB (mostly used in Korea), ISDB-T (in Japan), STiMi (China) and MediaFlo (mostly used in the USA).[3] Other regions, such as Latin America, have not yet opted for a concrete standard. To make the picture more complex, since 2008 it is technically possible for mobile devices to receive the DVB-T signal (the standard for DTT – Digital Terrestrial Television) and new standards are scheduled for the short term such as DVB-NGH (to overcome the limitations of current DVB-H) or IMB (to integrate operations of broadcast and unicast spectrum).

In the case of mobile networks, operators are now at a crossroads of infrastructural expansion and innovation. In particular it is not clear which is to be the tempo in which new infrastructures will be deployed or how the investments are to be returned. Moreover, uncertainties also concern the eventual impact of the intensive use of video applications and services in these networks. According to Cisco (Grivet, 2009), the data traffic over mobile networks is to increase seventyfold between 2008 and 2017 – mostly due to video and in a similar process to that of fixed Internet. In fact, the limited capacity of mobile communication networks is one of the main arguments in favour of the broadcasting model. However, in a context of the exponential growth of mobile Internet demand, this point has also become the main argument for operators to influence regulators according to their interests, which, as we shall discuss later, may result in a change in the conditions of both the Internet and the television ecosystem.

Economic Barriers

In the economic sphere, the main problem is to define a viable business model that fits the interests of all the actors involved. Up to now, the solution has been hanging over a triple revenue model: subscription fees, pay per view (or per download) and advertising. As in the other cases of emergent mobile contents (Feijóo, Gómez-Barroso and Martínez, 2010), advertising is the dominant revenue model. With a predicted total income of 10 billion dollars for mobile television in 2013, almost three quarters are to come from mobile advertising, with an average yearly growth of 20 to 25%. The contrast with the apparent failure of subscription (which is the prevalent option in broadcasting mobile television models) is evident: in 2009 there were only 3.2 million mobile TV users through subscription

3 A comprehensive comparison of the diverse mobile television standards can be found in Zhou et al. (2009).

fees. Japan, with the most successful broadcasting mobile tele-vision, deploys free-to-air mobile television with advertising as the main revenue source. Advertising is also a relevant option for Internet video (whether mobile or not) and companies like YouTube are trying to transfer their emerging business models to the mobile environment (Feijóo, Pascu, Misuraca and Lusoli, 2009). But even in the case of Japan and Korea, advertising is still not sufficient to ensure the profitability of broadcasting models, and new forms of combining and diversifying revenue sources are necessary (Enter/Idate, 2010).

The model of application and content stores (like AppStore by Apple or Android Market by Google) introduces a new scheme that is able to combine pay per download, advertising, and new income channels (such as in-app content commercialisation). However, the presence of television or video oriented applications is up to now limited to bridging functions that facilitate access to given contents and further possibilities remain to be explored.

Some of the mobile and software actors are also trying a hybrid model between application stores and IPTV. Apple's AppleTV and Google's Google TV constitute different attempts to transfer the logics of application stores to an eventual content store (again, Apple focusing on a walled-garden model and Google monetising searches and advertising). While this concerns mainly the hybridisation of Internet video and television content rather than mobile television, the long term strategy points to a convergent television and video content available on mobile devices (smartphones, tablets, netbooks) as well as at home (PC, television set).

Another relevant factor in making the business model definition more difficult involves the problems concerning access to copyright licences and their protection (Screen Digest, CMS Hasche Sigle, Goldmedia and Rightscom, 2006). It is not easy to conciliate the interests of companies that need to offer very attractive contents at a low price with the interests of distribution companies that demand additional costs to compensate for the risk of letting contents enter a new channel.

Normative Barriers

In mobile television services two different bodies of law collide (each with its own complexity): that of electronic communications (including spectrum policies) and that of audiovisual broadcasting regulation (including partially digital media). The convergence between these two worlds is still a pending matter.

Since mobile television is fundamentally an audiovisual service, the question of how to transfer those normative conditions operating in other audiovisual environments to the mobile environment needs to be considered: the protection of fundamental values, the supervision of pluralism in media, the limits to advertising practices and the promotion of cultural diversity may be crucial in the social and economic viability of this service (Bria, Kärrberg and Andersson, 2007).

Some of the more recent regulations in Europe are starting to include considerations along that line: The Law of Audiovisual Communications in Spain (Law 7/2010 from 31 March), for instance, explicitly refers to mobile television (art. 34), promoting the adaptation of contents to the specificity of the mobile platform and regulating the standards backed by the European Commission (DVB-H). In its first chapter, it extends the mandatory nature of plurality (political, ideological, cultural and linguistic) to the complete audiovisual sector, including mobile television. However, beyond these incipient steps, normative convergence remains a key challenge in Europe.

Mobile television is, however, a telecommunications service as well. For this reason it is necessary to resolve the conditions under which different actors may enter the market and the rules within it. Similar to the normative role in audiovisual broadcasting, the role of regulators here goes beyond that of a neutral referee, since their decisions may push the market in one direction or another, as, for instance, in the case of what is known as the "digital dividend" in Spain. The "digital dividend" refers to the part of the radio-electric spectrum that is freed after the process of digitalisation of the terrestrial television signal. It obviously arouses the interest of almost any kind of company using the radio-electric space. Television channels put pressure on keeping its use as a television spectrum, promoting an increase in the number of DTT channels. Mobile communication operators advocate assigning this free spectrum to the development of mobile broadband services. If, as has happened in other countries in Europe, the final solution for the distribution of the "digital dividend" is public auction, then the shape of the future mobile broadband services (including mobile video and television) depends on the bidding capacity of the mobile actors involved (Ramos, Pérez and Mascarell, 2007). Again, regional differences in the spectrum availability may arise as a consequence of the diverse distribution systems applied in different countries as a solution for the "digital dividend".

User Related Barriers

The obstacles for the development of mobile television services are not only on the supply side. One of the keys to understanding the difficulties in mobile television's take-off is on the side of the demand, specifically in the separation between the opinions and actual behaviours of users (Schuurman et al., 2009; Enter/Idate, 2010).

Stating that the problems in consolidating mobile television supply relate to a lack of attractiveness for users would be an oversimplification. In northern Europe, for instance, research shows a relevant interest in mobile television already in 2008: in Sweden 33% of mobile users were interested in mobile television (Westlund, 2008), and 16% in Finland (Verkasalo, 2008). However, these same surveys show a low percentage of actual adoption: 8% in Sweden and only 3% in Finland. The figures are significantly similar (or even lower) in the rest of Europe. A 2010 survey of 2,463 users of YouGovt in the UK claimed that 60% of the user sample identified at least one environment (while on a train, queuing, at work) in which they would be likely to use a free-to-air mobile television service. Users' attitudes therefore conciliate an apparently positive perception of the service with an extremely low adoption rate. Though the price – or more precisely, the cost perception – seems to be a relevant variable in order to explain this, some experiences may deny that assumption. The experience of the BBC in the United Kingdom, launching an experimental free service from May 2007 to April 2008 was revealing: they only reached a 600 person maximum audience per day and an average consumption of 13 minutes per month.[4]

The case of Spain is not dissimilar. In a survey on *Consumers and Convergence* carried out in 2010 by KPMG Consulting, from a sample of 5,622 mobile Internet users, 29% were willing to pay for contents. However, regional differences can be observed with regard to some countries in Europe: in the Netherlands only 6% would be willing to pay for contents, 12% in Ireland and 17% in Germany.

These figures show that the disagreement between demand and supply may respond to more complex causes than the mere adaptation of prices. Obviously programming is a relevant variable. On the basis of a study developed in Finland, Carlsson and Walden (2007) describe the average program profile according to users: short, easy to assimilate content of around 10 minutes (news, sports, cartoons, documentaries, etc.). It should never last more than 45 minutes (series) and it should be ready to be stored in the device, so that it can be accessed anywhere and at any moment. These characteristics make mobile TV consumption closer to that of Web TV rather than the standard programming model implemented to date.

The question concerning the moment of consumption is also a relevant one. Vangenck et al. (2008) describe in this respect three simulations carried out in Europe: in Helsinki users watched mobile television in commuting contexts, though connection at home (or at the workplace) was surprisingly high; in some of the big cities in Spain (Madrid, Barcelona, Seville), consumption at home was equivalent to that of commuting; and in Oxford, commuting was the preferred situation (37%), closely followed by the home (32%) and the workplace (23%). In all these simulations the prevailing reasons for mobile television consumption were "killing time" and "keeping oneself informed".

Outside Europe, the case of South Korea deserves a mention. In this country broadcasting mobile television (free-to-air and subscription based) coexists with

4 Online-Document: http://www.pocket-lint.com/news/16463/bbc-mobile-tv-trial-flunks. The question of price in this experience, however, should be clarified: though access to the television service was free, it was developed though agreements with mobile operators. Consequently, though experimental users did not pay for the television content, they did pay for the data transfer, at least in the cases without flat-rate data transfer.

telecom network based services, mostly deployed via subscriptions and on pay per view basis. Surveys by Chan-Olmsted et al. (2008) show an average consumption of 2-3 days per week, 30 minutes per day. The social contexts of usage are as a second TV set at home and in commuting. However, 50% of informants were willing to use mobile TV services only if they were free and 68% refused to pay for it in any case. Interestingly, in this case with a reasonable development and diversity of the services, price and quality are highlighted as crucial variables (Choi et al., 2008), together with "perceived usefulness" (Jung et al., 2009).

There are, however, other perspectives that offer a deeper view of the possible reasons for the divorce between the perceptions and behaviour of users. Åkeson and Eriksson (2007) point to a lack of coordination between diffusion models and usage models. Carlsson et al. (2006) consider this to be a general problem of mobile content, and García-García et al. (2010) underline that the real question is about interaction, not consumption. The specific condition of the mobile device as a relational technology (Aguado and Martínez, 2009) seems to lie in the background of a necessary differentiation between video or TV usage parameters and mobile video/TV consumption. This amounts to the fact that neither the knowledge of broadcasting television nor that of mobile communications may be decisive in order to find the way to fit users' expectations (Feijóo, Gómez-Barroso and Ramos-Villaverde, 2010). More specifically, Carlsson et al. (2006) point out that the mobile device is functionally associated with the experience of accessing the others and with knowledge of the environment, and this does not fit well with the entertainment conception inherited from other media. In the case of the mobile, users have to find specific contexts (time, space, emotional situation) to use a given service.

In the same line of thinking, Schatz et al. (2007) speak of a "mobile social television" as a "form of communication that creates a shared experience of watching television". In this case it is the user that pushes the innovation. However, the way to translate this into a viable business model is still to be found (Nicolajsen and Sorensen, 2009). The fact that, for example, in Spain most of the consumed mobile video content is user generated (49% versus some 10% of audiovisual services) supports the conception of the mobile device as a relational technology (Fundación Telefónica, 2010).

Mobile Communication Networks in the Age of Screens Convergence

The picture resulting from the main hurdles for mobile television services highlights an increasing relevance of mobile communication networks and a much slower development of audiovisual broadcasting solutions. The growth of mobile Internet and the launch of new mobile devices more specifically oriented to multimedia consumption (such as tablets) puts Internet video and networks at the epicentre of a process where different screens (PC, TV set, netbook, smartphone, game console, tablet, etc.) converge.

In a fuzzy context like this, an alternative strategy may be to consider the foundations of a convergent environment in which polyvalent business models may work. The case of application and content stores constitutes a possible ground for that. Apple TV, a project reborn in 2010 under the success of the App Store, intends to bring together the models of closed IPTV and that of the application and content store (like iBook store or iTunes), with the aim of providing video content available on any of the screens accessed by the platform (iPod, iPhone, iPad, iMac, iTV, etc). Similar initiatives, however with a much smaller range, are also being run by other actors, like hardware manufacturers (e.g. Samsung Movies). It is important here to consider the fact that audiovisual content producers are also looking for a viable model of distribution in the Internet, and that agreements with network operators that guarantee control over audience metrics and content licences become increasingly interesting (Montalvo, 2010).

The recent move by Google with the Google TV initiative should also be framed in this convergent landscape, although from a different point of view. The idea behind Google TV is to extend Google's search based revenue model to the audiovisual sector. To bring Internet video content (and Internet video search) to TV sets may mean a significant step in the integration of television and Internet. The evolution of software platforms (upon which content access and revenue models directly depend) points exactly to this screen convergence.[5] Content stores are to be multi-platform just as audiovisual contents tend to be multi-screen.

The increasing relevance of network operators in the current transformations of the television ecosystem also means a second chance for telcos to get their part of the digital content pie and to avoid being condemned to the dumb pipe role, as they were in the fixed networks. For this reason network operators have been consistently lobbying over the last year in two directions: to gain participation in the digital content industries, associating their revenues to the use of the networks, and to put pressure on to abandon the net neutrality principle.[6] The latter point especially involves the mobile ecosystem. Note that net neutrality is not fully operative in the mobile environment. For instance, in order to rationalize the increasing demand of mobile data traffic and the lack of bandwidth resources, operators may prioritise voice over data, including different user fees for different service

5 Apple's iOS version for tablets is clearly convergent in features and design with the ultimate version of Apple's PC operative system (OS X Lion), and Google's Android Honeycomb (the ultimate version of Googlegs mobile OS) is also aspiring to become a multi-device OS.

6 Net neutrality (or network neutrality) is a principle related to users' access to Internet networks that advocates no discrimination of the data traffic according to the origin (Internet providers' friend or rival companies) or the nature of contents, e.g. video, email, etc. (Ofcom, 2010).

level agreements. As a consequence mobile network operators might discriminate video oriented Internet access tariffs fro1m productivity oriented Internet use, changing *de facto* the very nature of the Internet environment and putting network operators in a strategic situation to take advantage of the screen convergence (Castellet and Aguado, 2010).

Conclusions

The future landscape of mobile television is all but clear. The main driver for its development lies in the rich video experience already available to mobile devices. The uncertainties surrounding the two competing models are not facilitating an appropriate orientation of the potential demand by the users.

The model based on audiovisual broadcasting networks is still at an early stage of commercial, legal and even technical deployment: there are no agreed standards or developed mobile broadcast networks and regulatory problems concerning spectrum licences are far from solved. The model based on mobile communication networks seems to be currently the more viable one. Currently available mobile television, at least in Europe, is in fact mobile network based.

However, actors from the mobile sector – especially operators – see mobile television as a secondary matter, basically useful for improving the attractiveness of their broadband services. The lack of involvement by operators in the consolidation of mobile television also concerns a strategy of rationalization of the increasing demand of mobile data traffic.

The changing horizon of mobile television services is not independent from the transformations that characterize both the television ecosystem and the digital content industries. The increasing demand of mobile Internet access and screen convergence in digital video put mobile networks in a privileged situation to demand regulatory and political support in order to enhance their capacity. The pressure mobile companies put on other players in exchange for the availability of their networks for the diffusion of television and video contents also increases. All these efforts converge in hinting at the current structure of mobile data access (including mobile television) as a reference for future Internet-based video networks.

Furthermore, the problems that mobile television must face in the coming years are similar to those concerning emergent forms of television: the standardization of the technical framework, the definition of a viable business model or the integration of audiovisual regulation with new technologies and consumption habits are all pending challenges in many of the faces of the screen convergence polyhedron.

The development of mobile devices, the enhancement of mobile networks and the consolidation of distribution platforms and innovative revenue models, together with the consolidation of mobile advertising constitute an opportunity to overcome most of the identified barriers for mobile television. Undoubtedly institutional commitment and, especially, regulatory support will be needed.

However, the most relevant challenge to be faced is the gap between product conception and users' perception. To presuppose that an attractive product sells itself automatically is an oversimplification that may drive mobile television to the margins of the mobile content business. The actors involved need to understand what the specificity of the mobile device involves for users, the usefulness they perceive in mobile television and how and in what contexts they intend to use it. Mobile television, thus, must be "mobilized" not only in technological terms, but specifically in the sense of adapting to the social meaning users give to mobility.

References

Aguado J. M., Martínez I. J. 2009. Mobile Media Implicit Cultures: Towards a Characterization of Mobile Entertainment and Advertising in Digital Convergence Landscape. Observatorio OBS*, 3 (1). URL: http://obs.obercom.pt/index.php/obs/article/view/253 (accessed 23 January 2012).

Åkesson M., Eriksson C. 2007. The Vision of Ubiquitous Media Services: How Close Are We? In: Human Interface and the Management of Information. Interacting in Information Environments. Berlin et al.: Springer.

Babe R. 1995. Communication and the Transformation of Economics: Essays in Information, Public Policy, and Political Economy. Boulder, Colorado: Westview.

Ballon P. 2007. Business Modelling Revisited: the Configuration of Control and Value. Info 9 (5), pp. 6-19.

Ballon P. 2009. The Platformisation of the European Mobile Industry. Communication & Strategies, 75 (3) quarter, pp. 15-34.

Bouwman H. 2003. State of the Art on Business Models. Enschede, Telematica Instituut.

Braet O., Ballon P. 2008. Cooperation Models for Mobile Television in Europe. Telematics and Informatics, 25 (3), pp. 216-236.

Bria A., Kärrberg P., Andersson P. 2007. TV in the Mobile or TV for the Mobile: Challenges and Changing Value Chains. Proc of Pimrc 2007. IEEE 18th International Symposium on Personal, Indoor and Mobile Radio Communications.

Carlsson C., Carlsson J., Puhakainen J., Walden P. 2006. Nice Mobile Services do not Fly. Observations of Mobile Services and the Finnish Consumers. eConference, eValues.

Carlsson C., Walden P. 2007. Mobile TV – To live or Die by Content. Proceedings of Hicss 2007, 40th Annual Hawaii International Conference on System Sciences.

Carlsson C., Walden P., Bowman H. 2006. Adoption of 3G+ services in Finland. International Journal of Mobile Communications, 4 (4), pp. 348-361.

Castellet A., Aguado J. M. 2010. Neutralidad de la Red y Contenidos Digitales: Conflictos de Intereses y Escenarios de Futuro. XXV International Conference CICOM 2010. Business Models for a Digital Economy: The Value of Contents. University of Navarre, 25-26 November.

CMT. 2010. Informe anual 2009. Barcelona: Comisión del Mercado de las Telecomunicaciones. URL: http://www.cmt.es/es/publicaciones/anexos/20100705_IA09_CMT_INFORME_ANUAL_2009_SENCER_BAIXA.pdf (accessed 23 January 2012).

Chan-Olmsted S. M., Lee S., Heo J. 2008. Developing a Mobile Television Market: Lessons from the World´s Leading Mobile Economy – South Korea. 8th World Media Economics and Management Conference.

Choi J. Y., Koh D., Lee J. 2008. Ex-ante Simulation of Sobile TV Market Based on Consumers' Preference Data. Technological Forecasting and Social Change, 75 (7), pp. 1043-1053.

Curwen P., Whalley J. 2008. Mobile Television: Technological and Regulatory Issues, Info, 10 (1), pp. 40-64.

DVB-H Forum. 2007. URL: http://www.dvb-h.org/faq.htm#11 (accessed 23 January 2012).

Enter/Idate, 2010. Mobile, 2009: Market & Trends, Facts & Figures. Instituto de Empresa: Madrid.

Feijóo C., Gómez-Barroso J. L. 2009. Factores clave en el acceso móvil a contenidos. El profesional de la información, 18 (2), pp. 145-154.

Feijóo C., Gómez-Barroso J. L., Martínez-Martínez I. J. 2010. Nuevas vías para la comunicación empresarial: publicidad en el móvil. El profesional de la información, 19 (2), pp. 140-148.

Feijóo C., Gómez-Barroso J. L., Ramos-Villaverde S. 2010. Medios de comunicación en Internet móvil: la televisión como un modelo pendiente de éxito. El Profesional de la Información, 19 (6), pp. 637-644.

Feijóo C., Maghiros I., Abadie F., Gomez-Barroso J. 2009. Exploring a Heterogeneous and Fragmented Digital Ecosystem: Mobile Content. Telematics & Informatics, 26 (3), pp. 282-292.

Feijóo C., Pascu C., Misuraca G., Lusoli W. 2009. The Next Paradigm Shift in the Mobile Ecosystem: Mobile Social Computing and the Increasing Relevance of Users. Communications & Strategies, 75 (3), pp. 57-78.

Freeman C., Soete L. 1997. The Economics of Industrial Innovation. Cambridge, MA: MIT Press.

Fransman M. 2007. The New ICT Ecosystem: Implications for Europe. Edinburg: Kokoro.

Fundación Telefónica. 2010. Informe de la sociedad de la información en España. Madrid. URL: http://e-libros.fundacion.telefonica.com/sie10 (accessed 23 January 2012).

García-García A.,Vinader-Segura R., Albuin-Vences N. 2010. Televisión tradicional y televisión móvil. Estrategias para contenidos televisivos en movilidad. Telos, 83, pp. 84-96.

Grivet V. 2009. Business Models in Mobile TV. Paris Tech Telecom. URL: http://innovation-regulation2.telecom-paristech.fr/wp-content/uploads/Documen-ts/thematiques/Innovative_Business_Models/Grivet_18_12_09_V2.pdf (accessed 23 January 2012).

GSA. GSM/3G market update, 2010. URL: http://www.gsacom.com/downloads/pdf/GSM_3G_Market_Update.php4 (accessed 23 January 2012).

Hee Shin D. 2006. Prospectus of Mobile Television: Another Bubble or Killer Application? Telematics and Informatics, 23 (4), pp. 253-270.

Iansiti M., Levien R. 2004. The Keystone Advantage – What the New Dynamics of Business Ecosystems Mean For Strategy, Innovation and Sustainability. Harvard: Harvard Business School Press.

Jung Y., Pérez-Mira B., Wiley-Patton S. 2009. Consumer Adoption of Mobile TV: Examining Psychological Flow and Media Content. Computers in Human Behavior, 25 (1), pp. 123-129.

Kretzschmar S. 2006. Journalistic Content and the Football World Championship 2006: Multimedia Services on Mobile Devices. 2006 ICA Pre-Conference "After the Mobile Phone?", 16-18 June. University of Erfurt, Germany.

Montalvo J. 2010. Las televisiones buscan sus pantallas. La televisión que viene. Cuadernos de Comunicación EVOCA. Madrid, pp. 35-40.

Nicolajsen H. W., Sørensen L. T. 2009. User as Innovators – User Needs for Future Converged Mobile Media Services. In: Proceedings of the 32nd Information Systems Research Seminar in Scandinavia, Molde University College, pp. 567-581.

Ofcom. 2010. Traffic Management and 'Net Neutrality'. A Discussion Document. URL: http://stakeholders.ofcom.org.uk/binaries/consultations/net-neutrality/summary/netneutrality.pdf (accessed 23 January 2012).

Prario B. 2007. Mobile Television in Italy: Value Chains and Business Models of Telecommunications Operators. Journal of Media Business Studies, 4 (1), pp. 1-19.

Ramos-Villaverde S., Pérez J., Mascarell B. 2007. Mobile TV Services Provision Scenarios – Implications of Spectrum and Audiovisual Policies in the Development of Mobile TV Market in Europe. The Journal of the Communications Network, 6 (1), pp. 40-47.

Schatz R., Wagner S., Egger S., Jordan N. 2007. Mobile TV Becomes Social – Integrating Content with Communications. In: Proc of ITI 2007, 29th International Conference on Information Technology Interfaces, Zagreb (Croatia), pp. 263-270.

Schuurman D., De Marez L., Veevaete P., Evens T. 2009. Content and Context for Mobile Television: Integrating Trial, Expert and User Findings. Telematics and Informatics, 26 (2), pp. 293-305.

Screen Digest, CMS Hasche Sigle, Goldmedia and Rightscom. Interactive Content and Convergence: Implications for the Information Society. London: European Commission. DG Information Society and Media (S. Digest).

Uglow S. 2007. The Race for Mobile Content Revenues. Juniper Research.

Vangenck M., Jacobs A., Lievens B., Vanhengel E., Pierson J. 2008. Does Mobile Television Challenge the Dimension of Viewing Television? An Explorative Research on Time, Place and Social Context of the Use of Mobile Television Content. Changing Television Environments. Lecture Notes in Computer Science, Vol. 5066/2008. Berlin: Springer, pp. 122-127.

Verkasalo H. 2008. From Intentions to Active Usage: a Study on Mobile Services in Finland. In: 19th European Regional Conference of the International Telecommunications Society.

Westlund O. 2008. Diffusion of Internet for Mobile Devices in Sweden. Nordic and Baltic Journal of Information and Communications Technologies, 2 (1), pp. 39-47.

Zhou J., Ou Z., Rautiainen M., Koskela T., Ylianttila M. 2009. Digital Television for Mobile Devices. IEEE multimedia, 16 (1), pp. 60-70.

Andrea Miconi

A Glocal Way to Broadcasting: Neighbourhood TV and Web TV in Contemporary Italy

Introduction

This chapter aims to identify the specific role played by a particular actor – the neighbourhood TV stations – from their origins to their evolution into Web TV stations. It draws on current literature on the transition to new media. The Italian debate on the subject is characterised by two opposing traditions, with the mere description of media market trends on the one side, and the ideological-critical interpretation of the phenomenon on the other. In the latter case, the rise of new TV stations is perceived to be the free expression of new voices coming from the grassroots, previously hindered by the television monopoly. In this sense, the paper tries to focus on the contribution of amateurish TV stations, and describes their role in the Italian broadcasting system by the following three steps: first, a brief assessment of the historical context that has produced such experiences; then, a map of the phenomenon of neighbourhood TV stations, with particular reference to the most successful ones, and to their ability to provide a new account of the territory; finally, an analysis of the transformation of these stations into Web TV stations, and of the opportunities provided by the implementation of new technologies.

Historically, Italian television broadcasting has always been a battlefield and a competitive arena for the great powers: first throughout the period of the state monopoly on broadcasts (1954-1975), then during the deregulation phase (1975-1983), which followed the emergence of new private broadcasting *lobbies* (Forgacs, 1990, p. 142) and resulted in the RAI/Mediaset duopoly, and finally with the advent of the digital platforms taken over by the Sky tycoon, Rupert Murdoch, in 2003. In such a scenario, the introduction of digital terrestrial television by decree law, with the so-called "Gasparri Law" passed in Italy in 2004 – an unusual case of innovation driven by bureaucratic and institutional means – seems to mark the latest chapter in a story entirely dominated by the intertwining of public and private affairs, overwhelmed by the pressure of the great powers, and scarcely open to the cultural demands of the grassroots. In short, such a succession of management models has never actually questioned the centrality of television in the Italian cultural industry, nor its anomalous semi-monopolistic control, which is moving further into the third millennium and into the age of *new media* (Ortoleva, 2005, pp. 285-286).

A first potential challenge to the monopolistic or semi-monopolistic structure of the market was actually posed by the private Hertzian television networks, which suddenly emerged in the second half of the Seventies, and usually broadcasted in very narrow areas, giving voice to local values and needs, but clearly reflecting, at the same time, the entrepreneurial weakness of the Italian territorial capitalism (Tinagli, 2008, pp. 132-133). While contributing to the quantitative expansion of the market, local networks were not, however, able to question the leadership of major groups: on the contrary, deeply depending on the advertising revenue, they ended up producing a sort of paradoxical enhancement and legitimation of the private/public duopoly (Richeri, 2009, p. 484). If we look at the overall statistics of this period, about 30% of the advertising revenues were collected by Rai TV, while the national private broadcasters attracted 60%, and hundreds of local networks collected no more than 9% (Richeri, 2009, p. 485). So, in an advertising-driven market, their economic weakness prevented local networks from emerging as competitive subjects in the cultural industry. On the contrary, since it became available with the advent of digital media, the streaming transmission marked a real point of discontinuity (De Rosa, 2001, pp. 1-3), enhancing the long tail of distribution (Anderson, 2004) and the search for niche local markets. The increased difficulty in the audience measurement, in this sense, allowed Web TV to establish a different model, less depending on advertising (De Rosa, 2001, pp. 25-29), and destined to attract and collect all the social and cultural experiments in local broadcasting. The domestication of Web TV and its increasing appearance in everyday life, in this sense, forced the Italian market towards a new organization of both old and new media (Cola, Prario and Richeri, 2010, p. 96), leading the system to eventually overcome traditional powers and well-established alliances (Carnevale Maffé, 2011, p. 185).

Yet, in the same years when Mediaset and Sky engaged in a larger battle between national and global sovereignties, a number of smaller and more dynamic players timidly started to emerge on the opposite front: the so-called *telestreets*, or neighbourhood TV channels.

"Micro-Broadcasters"

Termini Imerese is a small town on the North coast of Sicily, heading east from Palermo, just outside the orange grove known as "Conca d'Oro" (The Golden Shell). However, it has almost exclusively hit the headlines because of the FIAT factory, set up in the early 1970s, specialising in the production of low-powered cars in compliance with the development agreements for the Southern regions. On top of that, Termini Imerese is also known for the social upheaval caused by the progressive dismantling of the plant, and by the crisis that has gradually reduced the number of people in employment. In autumn 2002, after the umpteenth "zero-hour" wages guarantee fund – a typical Italian policy response to the economic crisis – some of the local workers were on strike and in search of a means to voice their protest. Some of them probably happened to read *The Archipelago of Ethereal Shadows*, the manifesto of Orfeo *Telestreet*, which had established itself in Bologna, at 29 via Rialto, a few months before, on the day of the summer solstice:

- Orfeo is a micro-broadcaster, a street television channel.
- Orfeo covers an area of several hundred metres.
- Orfeo is the self-funded, voluntary fruit of a constantly evolving common sensitivity.
- Orfeo knows it is punishable by law.
- Orfeo knows it can rely on Article 21 of the Constitution.

The idea of Orfeo was strikingly simple: in order to express a viewpoint not acknowledged by the broadcasting media, it was necessary to start broadcasting autonomously, taking advantage of the spaces made available by the failures of technology and the legislative void. The contradiction between the last two points of Orfeo's manifesto is basically the core of the problem raised by the participatory experience of *telestreets*: while drawing on freedom of expression, and on Article 21 of the Italian Constitution[1], street TV stations were nevertheless forced to broadcast illegally. More precisely, *telestreets* were able to transmit their signals by occupying the so-called "shadow cones", i.e. the frequencies allocated to the great TV networks, but unusable because of territorial obstacles: in the city of Bologna, it was frequency 51, which had been allocated to MTV Italy, but was, nevertheless, free in the neighbourhood, whereas in the town of Termini Imerese it was frequency 31. By means of low-cost transmitters operating within a radius of a few hundred metres, *telestreets* thus filled the voids existing in the media industry, taking advantage of its technical faults and hidden pockets, and writing a new chapter in the struggle for the appropriation of cultural spaces, which has its

1 Article 21 reads as follows: "All have the right to express freely their own thoughts by word, in writing and by all other means of communication. The press cannot be subjected to authorisation or censorship. Seizure is permitted only by a detailed warrant from the judicial authority in the case of offences for which the law governing the press expressly authorises, or in the case of violation of the provisions prescribed by law for the disclosure of the responsible parties. In such cases, when there is absolute urgency and when the timely intervention of the judicial authority is not possible, periodical publications may be seized by officers of the criminal police, who must immediately, and never after more than twenty-four hours, report the matter to the judicial authority. If the latter does not ratify the act in the twenty-four hours following, the seizure is understood to be withdrawn and null and void. The law may establish, by means of general provisions, that the financial sources of the periodical press be disclosed. Printed publications, shows and other displays contrary to morality are forbidden. The law establishes appropriate means for preventing and suppressing all violations".

roots in the movements of 1977 and in the counter-information battles (Berardi, Jacquemet and Vitali, 2003, pp. 128-132).

This battle was, however, bound to end with some defeats, as in the case of Telefabbrica, which briefly reported on the FIAT workers' strikes, before being shut down by the Ministry of Communications at the beginning of December 2002. Set up by a group of disabled people in Senigallia, in the Marche region, Disco Volante neighbourhood TV would also suffer the same fate, in spite of being supported by the local administration, and was shut down in 2004. Nevertheless, having addressed some crucial issues, such as the working-class struggle, neighbourhood TV stations gained an unprecedented visibility, marking a turning point, or at least an important step, in the history of Italian communication, and triggering a long series of similar initiatives. Among the most well-known stations, UniversiTV was set up soon afterwards, in 2005, at the Third University of Rome, with the purpose of reporting on the university's internal affairs, under the direct management of the students from the Faculty of Communication.

Right from the beginning, the peculiarity of Italian street TV was to operate within a particularly limited, and deliberately local, area. Far from being institutional organisations, such as the Community TV channels set up in Manhattan and diffused in many countries, the Italian neighbourhood TV stations immediately established themselves as "micro-TV's": they were run quite amateurishly and, precisely for this reason, were able to drastically reverse the traditional top-down mass communication model (Dagnino and Gulmanelli, 2005, pp. 70-71). Since 2002, dozens of street-TV stations have been founded, even if, according to some, the first experiments should be traced back to TeleMonteOrlando, which started its transmissions in Gaeta on Christmas Eve 2001. Given their informal and amateurish nature, they constitute an archipelago of different experiences, a polyphony of interests and voices, a joyful chaos machine, able to shake, at least for a time, the traditional structure of the Italian communication system.

As for their content, the experiences of micro-TV stations seem to obey to different missions. If we look at the classical McLaughlin's economic model, in particular, media are supposed to develop their business model according to a four-dimensional pattern, characterised by the oppositions between product and service, on the one hand, and devices and contents, on the other (McLaughlin and Birinyi, 1980). Regarded as one of the most suitable models for framing the history of Italian communication (Gambaro, 1985; Ortoleva, 1994, pp. 28-31), this model can easily be applied to our case, with the following categorisation of the micro-TV sector:

First of all, we have the Web-TV stations focused on the dimensions of devices and contents, or, in other words, mainly interested in the diffusion of new technological platforms for developing cultural and political issues. It is the case, for example, of the strictly antagonistic TV stations, inspired by the principles of *media activism*, and representing in some way the Italian version of the counter-cultural practices established in most parts of the world, and inspired by Seattle's waves of protest, *no-global*, and later *new-global* movements. They started taking shape after the experience of Orfeo in Bologna and Candida in Rome. An emblematic moment in their story was the support given by dozens of Italian street TV stations to the demonstration against the war in Iraq on 22 February 2003, which, on the one side, represented the first agreement and integration among different micro-TV stations[2], and, on the other, combined the media activism practice with the rainbow flags of the peace protest and the anti-militarist movement of opinion (Veltri, 2005, p. 48);

Then we have the micro-TV stations focused on the dimensions of contents and services, being aimed at adding some social instances to the public debate. In this case, Neighbourhood TV stations were created to voice the viewpoints of small communities on local issues, such as TeleMonteOrlando in Gaeta, Torre Maura Television in Rome, and TeleStreet Bari. In the large cities, in particular, street TV stations have tried to give accounts of the daily life in disadvantaged urban areas, as in the well-known case of Insu^TV in Naples, with its reportages from Scampia, the district made famous by Roberto Saviano's novel *Gomorra*. Again, some street-TV experiments were promoted with the direct participation of institutions, with the purpose of mobilising the positive energies of the suburbs. Among them, the most cited case is represented by the Corviale Network, an integrated project promoted in a disadvantaged area of Rome by left-wing councillor Luigi Nieri, who also set up a micro-TV channel able to broadcast on different platforms, from frequency 31 of the terrestrial signal to satellite and the Sky package.

Finally, on the opposite side, we have micro-TV opting for the dimensions of content and product, such as in some extemporaneous broadcasting experiments taking advantage of the shadow cones of the official frequencies, and using the traditional technical equipment for piracy purposes. In this respect, a well-known episode was the transmission of the football match Juventus – Roma on channel 26, on 21 September 2003, in the popular Roman district of San Lorenzo, which was intentionally organised as a violation of the Italian football television rights (Patanè Garsia, 2004, p. 238).

---

2 Besides the political significance of the initiative, the organisation of a common transmission represented a moment of convergence among various neighbourhood TVs, with interesting implications for the whole communication system. The TV stations participating in the protest of 22 February 2003, were: Pitbull TV, Tele Aut-TeleStreet, Challenger Telestreet and SpegniLaTV-Telestreet from Rome, NomadeTV and MosaicoTV from Milan, Albornoz, TeleImmagini and TVTB-Telestreet from Bologna, RagnaTele, Este TV and Challenger TV from Padua, La confraternita del rosso (Monopoli, Bari), Gli anelli mancanti TV (Florence), La voce del Sud (Gaeta, Latina), Tele In (Scauri, Latina), TiVitti (Palermo), Ottolina TV (Pisa), SienaCrew (Siena), NO Privilegi Politici (Vicenza). On that occasion, Telefabbrica also resumed its broadcasts from Termini Imerese.

However, such a constellation of interests did not prevent micro-TV stations from teaming up and merging into a network, which was eventually bound to constitute a proper *player*: whereas the word *telestreet*, with a lower-case "t", actually refers to single broadcasting experiences, Telestreet, with a capital "T", more properly indicates the *network* that collects them, in an attempt to organise and rationalise their efforts (Bazzichelli, 2008, p. 255). Indeed, it is here that the decisive game for the future of street TV stations unfolds: the search for a final legitimisation, the institutional acknowledgement of their role, along with regulation of the sector, which the Gasparri Law failed to provide.

According to many, the current experience of micro-TV stations recalls that of cable TV, which, in the early 1970s, first attacked the public broadcasting monopoly (starting from TeleBiella in 1971): those were the years when many small networks, run by amateurs and operating locally, started the "legal battle" over the acknowledgement of their own legitimacy (Monteleone, 1992, p. 386), and suddenly challenged the hegemony of public television in the national culture. There are actually some common features in both their stories: the questionable legitimacy of their transmissions, their local extent, and their tendency to federate into a more solid structure. As for the first aspect, it is no accident that both parties – local TV stations thirty years ago, and micro-TV stations today – appealed to Article 21 of the Italian Constitution, which inspired the well-known decision of the Constitutional Court (no. 202 of 28 July 1976) to authorise "the installation and use of over-the-air radio and television stations within a local area", thus paving the way for the mixed public/private media system and for a great change in the Italian cultural industry. As for the latter aspect, both private TV stations and micro-TV stations have tried to develop a larger organisational structure, in order to withstand the pressure of the market: the first through the establishment of a national circuit, the latter through the Telestreet network and a series of initiatives launched by the meetings called "Etera", such as those held in 2002 in Bologna, the city where Orfeo was first created, and in 2004 in Senigallia, another highly symbolic place due to the activities of Disco Volante (whose close-down was also disputed by many MPs, who decisively contributed to the emergence of a public debate over the issue).

Nevertheless, between the two experiences there are also some significant differences: micro-TV stations do not actually have a well-defined business plan, whereas the old private channels had a clear commercial purpose. Besides, micro-TV stations aim to serve as an interlocutor for both the audience and the institutions, thus assuming, at least implicitly, a political relevance (Berardi, Jacquemet and Vitali, 2003, p. 37). Moreover, the age of street TV is characterised by technological opportunities that were absolutely unthinkable thirty years ago, which makes it possible to combine the social mission of neighbourhood TV stations with the new framework of Web TV. Although there is yet no comprehensive and univocal definition of Web TV (Noam, 2004, p. 4), in some respects this phenomenon can be regarded as a *glocal* format intended to convey local issues through the possibilities offered by digital technologies (Pescio, 2007, p. 101). There is therefore a clear elective affinity between Web TV and the experience of micro-TV stations, which might provide grassroots movements with a new powerful weapon. In fact, in 2004, following the example of Insu^TV, the Linux-based SOMA platform started being widely used, making it possible to create an automated broadcasting scheduler and turn the amateurish experience of neighbourhood TV stations into an intensive exploitation of the opportunities provided by the Web. It is here, then, that a new chapter in this story begins.

From Neighbourhood TV to Web TV

"Millions of dwarfs" challenging the giants of communication: this is the image used by Michele Mezza (2011) and Derrick de Kerckhove (2011) to describe Web TV. Here we are facing the domestic version of that phenomenon of *participation* (Tapscott and Williams, 2006, p. 11) and "insurgent" communities of practice (Castells, 2009, pp. 299-415) typical of the digital age, with the pressures of global trends on the one side, and the restrictions imposed by some typically Italian peculiarities on the other.

As for the extent of such a phenomenon, statistics currently available provide contradictory information as often happens in suddenly rising markets: FEMI, the micro-TV Italian Federation, numbers at the moment 126 stations[3], while another network, the Italian Association of Web Television (AssoWebTV)[4], established in march 2008, is not yet able to provide any information, and according to other sources, again, in the early 2010 there were 163 stations (Coletti, 2010, pp. 179-183). On the other hand, the more ambitious Netizen report, updated to December 2010, gives a more optimistic picture of the situation (Altra TV, 2010): according to the data reported, in Italy there are some 436 micro-TV stations operating on the Web, an increase of more than 50% compared to the figures for 2009, when the total number was 286. The incidence of some geographical inequalities must certainly be taken into account, since not all regions share the same level of activity and initiative[5]; however, it is from a qualitative perspective that the Italian case reveals its most interesting peculiarities.

3 Data are available at www.femiTV.tv.

4 Online-Document: www.assowebTV.it.

5 Interestingly, according to the Netizen report, the line between active and inactive areas does not coincide with any traditional geographical divide: both Northern (Lombardia, Piemonte) and Southern regions (Campania, Sicily) feature among the most advanced promoters of this media activism practice. Analogously, the map of the excluded areas is also very irregular, including Apulia and Veneto, Valle d'Aosta and Calabria, Friuli Venezia Giulia and Liguria.

As Roland Robertson wrote some time ago, a crucial factor in understanding the overall process of globalisation is the interwoven relationship "between the particular and the communal, on the one hand, and the universal and the impersonal, on the other" (Robertson, 1992, p. 103). From that point of view, as mentioned before, micro-TV stations seem perfectly to embody the *glocal* nature of contemporary culture. As soon as neighbourhood broadcasters land on the Web, they end up establishing a virtuous relationship between their own local reach and the powerful means provided by new technologies: among the persons in charge of the 436 micro-TV stations surveyed by Netizen, in an attempt to compensate for the weaknesses of the great broadcasting companies, as many as 37% claim that their objective is "community information", while another 32% more explicitly indicate "local promotion" as their main objective, aiming to enhance the area where the broadcasting practice takes place (Altra TV, 2010). Over two-thirds of Web TV stations, therefore, claim to serve their own community, whereas only 5% of the interviewed point out "social criticism" as their own main goal, explicitly referring to more political objectives; the remaining 26%, finally, describe themselves as "thematic channels" providing supplementary information services in addition to the offerings of mainstream broadcasters. Browsing through the Netizen report – which provides an exhaustive view of the phenomenon, and from here onwards will be regarded as our main source of information – it is possible to see that, quite surprisingly, one-third (33%) of the Web-TV managers interviewed claim that they are not interested in having direct relations with the public administration, whereas others, in a similar proportion, aim to obtain at least some sort of official recognition (26%), if not some practical cooperation (34%), from local administrations, and only a small minority (7%) benefit from public funds in support of their activities. On the whole, most of these initiatives appear to be largely amateurish, spontaneous and autonomous, as they have been made possible by the easy access and low costs of the digital network. The financial aspect is going to play an increasingly crucial role, however, in the future development of micro-TV stations.

Currently, according to the Netizen report, as much as 62% of Italian micro-TV stations rely only on self-financing, with no external support whatsoever; 19% benefit from occasional private funds; less than 10% are supported, as already mentioned, by public administration; 8% take advantage of European funds; and only 2%, a marginal quote, are able to position themselves in the market, offering on-demand video services. In this respect, the Italian Web TV stations still fail to keep up with the Anglo-American ones, which, in the last few years, have registered significant increases in funds and investments (Lasica, 2005). This might depend on three factors. First of all, the Italian-language market is evidently smaller than the Anglo-American one, and, on top of that, the local character of the stations does not facilitate the diffusion of their transmissions at a national level. Secondly, even if Web platforms have favoured the proliferation of neighbourhood TV stations, the Italian delay in the diffusion of broadband and high-speed Internet access is a strong limitation to their development, representing a sort of "second-level" digital divide (Selwyn, 2004, pp. 351-354). Technology aside, a similar delay can be found in the circulation of Web culture and online practices in large sectors of the population, which are mostly penalised by low cultural capital (Sartori, 2006 pp. 101-103; Bentivegna, 2009, p. 67), and are instead sensitive to the traditional message of TV. Lastly, the militant background or inspiration of many neighbourhood TV stations – such as Orfeo and the great number of stations promoted by activists – has probably favoured their open participative nature, but has also negatively affected their ability to develop a proper market strategy and business plan. It is also necessary to understand whether it is only due to a *delay*, or – given the peculiar combination of backwardness and progress characterising the Italian media industry – whether it is rather the first actualisation of that sort of non-market-based, non-profitable rationality that many theorists see in action behind the collaborative logic of the Web (Anderson, 2004 p. 67; Benkler, 2006, pp. 3-4).

If we also examine the structure of the editorial staff of Italian Web TV stations, drawing on the Netizen report, we can find confirmation of their militant, participative, informal nature. Of the micro-TV stations, 62% have *micro* dimensions, with editorial teams made up of no more than 5 persons; 22% employ between 6 and 10 persons; and only 8%, a small minority, have more than 10 persons, and can therefore perform a proper journalistic division of labour. Web TV stations therefore have very *agile* editorial teams, mostly made up of young people: 34% of the workers are aged between 21 and 30; 44% are in the age range of 31 to 40; and only 4% are over 50 years old. Such a *young*, open, and *amateurish* nature of Web TV seems to form the bridge between the two phases of this story: on the one hand, it recalls the local dimension of the early analogue neighbourhood broadcasters, on the other hand it is also typical of the Web collaborative platforms, where people get themselves involved through informal practices and "self-selection" processes, so that, in digital work environments, one may happen to come into contact quite frequently with "totally unexpected partners", if not with accidental ones (Cottica, 2010, p. 50). However, this amateurish nature inevitably affects, and not painlessly, the organisation of the editorial work, because the "disintermediation" of broadcast media implies the risk of losing control over the procedures and quality of the final product (Rosenberry, 2010, p. 152). The search for a balance between an open collaborative model and rigorous work practices is a challenge that micro-TV stations cannot afford to ignore any longer.

A last remark regards the content offered by micro-TV stations, which is inevitably difficult to monitor, given the large diffusion of the phenomenon, their dispersion across the country, and the rather extemporaneous, non-scheduled nature of their transmissions. According to the Netizen report 2010, the offer of Italian micro-TV stations mainly consists of features (25%), interviews (20%), reportages (17%), news and strictly information services (16%), and documentary films (16%). Only a small percentage (6%) is made up of live streaming programmes, which confirms the Italian technological backwardness. The complete absence of entertainment programmes proves that the mission of micro-TV stations, first in analogue and then in digital format, remains that of providing a new local information channel, free from the influence of the great traditional broadcasters. In this respect, it would be nothing more than the Italian version of the great phenomenon of "citizen journalism", which is based on the "ordinary person's capacity to bear witness" to crisis events unfolding around them, such as the South Asian tsunami of 2004 (Allan, 2009, p. 18), and is here translated into a more ordinary, *daily*, local version. As already pointed out, the attention given by Web TV stations to local issues does not only depend on their information and social mission, but also, more prosaically, on the attempt to attract a critical mass of local audience unsatisfied by generalist media (Glaser, 2004). Clearly, the future of micro-TV stations also, and *above all*, depends on that. As for their success with the public, according to the Netizen report, 43% of Italian Web TV stations have less than 3,000 monthly unique visitors; 22% attract between 3,000 and 7,000; and 8% reach over 10,000. Whether they are *few* or *many*, it is not a matter of opinion but depends on the fulfilment of Chris Anderson's prophecy (Anderson, 2006, p. 23) about "the end of the blockbuster era" and the establishment of the long tail principle, according to which the sum of many small markets is worth as much as the great generalist information and entertainment market (Anderson, 2006, p. 45).

Discussion

In conclusion, the role of new technologies in turning micro-TV stations into central players in the Italian communication system can certainly be discussed, but is yet to be proved. It will only be possible to draw some conclusions in the next few years, when the gap that separates Italy from most Western European countries, with regard to the diffusion of Information and Communication Technologies and broadband services, is presumably bridged. In any case, new technologies seem to offer a possible, *latent* answer to a *real* problem, which has arisen with the loss of credibility of *mainstream* media in the eyes of a small, but ever-growing audience. In this regard, the success of "Rai Per Una Notte", an initiative promoted by the Italian National Press Federation (FNSI) on the night of 25 March 2010, is emblematic: on the occasion of the live streaming of the programme, which was organised as a protest against some increasing restrictions on the freedom of the press, many personalities banned from TV programmes for a long time (such as Sabina Guzzanti and Daniele Luttazzi) came back on stage. In the same way, the success of Current TV, Beppe Grillo's blog, and the newspaper *Il Fatto Quotidiano* (the first to refuse public funds for the press), proves that part of the audience is starting to have different needs from those detected by the great traditional media. In this respect, neighbourhood micro-TV stations, in their own small way, have been the first vehicle for such a change.

If we try to draw some general inferences, starting from the current scenario, we can outline three main considerations, which help us point out the most relevant questions.

First of all, transitional forms such as Web TV, typically resulting from a fusion or a "re-mediation" between the traditional audio-visual platforms and the digital devices (Fidler, 1997, pp. 232-233; Bolter and Grusin, 1999, pp. 233-237), can play a particularly relevant role in a market traditionally affected by a technological delay, and bring the public closer to the unfriendly new media.

In particular, and secondly, this gradual domestication of new technologies is, not surprisingly, performed through the enhancement of a local sense of community: in this sense, our evidence seems to confirm the close relationship between the proliferation of digital media and the empowerment of a local sense of belonging and identity, as shown in many recent studies (Hampton and Wellman, 2003, p. 277; Wellman, 2004, p. 28). If we look at the people interviewed in Web-TV programs, for example, we find that micro-TV stations, unlike traditional media, are not attracted to celebrities (27% of the overall number of guests), and usually prefer to involve common people living in the local area (57% of the total). Similarly, the most part of Web-TV stations, around 83%, are managed by inner community members (Colletti, 2010, p. 43).

So, in conclusion, are we talking about a real political revolution, as in the original *Telestreet* manifesto, made possible by the concurrence of digital devices, small territorial extension and decision-making practices (Gerodimos, 2004, p. 29)? Even if micro-TV stations were originally inspired by a conscious ideological project, as we have seen, cultural and social issues seem to have acquired a more decisive relevance in the period of their second and more significant proliferation. Cultural contents, in fact, represent an increasing part of the Web-TV programming, covering 22% of the overall offer (Colletti, 2009, p. 44), and we can find similar indications if we look at their "micro-zeitgeist", mostly dominated by issues related to "environment" and "job", contrary to the clear preference of traditional media for political debate (Colletti, 2009): in this sense, such a new peculiar independence of a social domain from politics seems to be the last contribution of micro-TV to the development of the Italian public opinion (Pasquinelli, 2002, p. 146).

References

Allan S. 2009. Histories of Citizen Journalism. In: S. Allan & E. Thorsen (eds.). Citizen Journalism: Global Perspectives. New York: Peter Lang, pp. 17-31.

Altra TV 2010. Netizen (Internet Citizen 2010). URL: http://www.altratv.tv/ (accessed 15 January 2012).
Anderson C. 2004. The Long Tail. London: Random House Business Books.
Bazzichelli T. 2008. Networking: the Art as Network. Aarhus: Digital Aesthetics Research Center.
Benkler Y. 2006. The Wealth of the Networks: How Social Production Transforms Markets and Freedom. New Haven et al.: Yale University Press.
Bentivegna S. 2009. Disuguaglianze digitali. Le nuove forme di esclusione nella società dell'informazione. Roma-Bari: Laterza.
Berardi F., Jacquemet M., Vitali G. 2003. Telestreet: macchina immaginativa non omologata. Milano: Baldini e Castoldi Dalai.
Bolter J. D., Grusin R. 1999. Remediation. Understanding New Media. Cambridge: MIT Press.
Carnevale Maffè, C.A. 2011. Eccola, dunque, la rivoluzione. Il software entra in TV (con Google, Apple & Co.). Link, 10, pp. 177-185.
Castells M. 2009. Communication Power. Oxford et al.: Oxford University Press.
Cola M., Prario B., Richeri G. 2010. Media, tecnologie e vita quotidiana: la domestication. Roma: Carocci.
Colletti G. 2009. L'ambiente fa notizia. Nova, December 15th 2009.
Colletti G. 2010. TV fai-da-Web. Storie italiane di micro Web TV. Milano: Il Sole 24 Ore.
Cottica A. 2010. Wikicrazia. L'azione di governo al tempo della rete: capirla, progettarla, viverla da protagonista. Palermo: Navarra Editore.
Dagnino A., Gulmanelli S. 2005. PopWar. Il NetAttivismo contro l'ordine costituito. Milano: Apogeo.
De Kerckhove D. 2011, Introduzione a M. Mezza, 2011, Sono le news, bellezza! Vincitori e vinti nella guerra della velocità digitale. Roma: Donzelli.
De Rosa J. 2001. Web Television. Fare la TV su Internet. Faenza: Editrice Cinetecnica.
Fidler R. 1997. Mediamorphosis. Understanding New Media. Thousand Oaks: Pine Forge Press.
Forgacs F. 1990. Italian Culture in the Industrial Era (1880-1990). Manchester: Manchester University Press.
Gambaro M. 1985. Informazione, mass media e telematica. Milano: Clued.
Gerodimos R. 2004. Democracy and the Internet: Access, Engagement and Deliberation. Journal of Systemics, Cybernetics and Informatics, 3 (6), pp. 26-31.
Glaser M. 2004. The New Voices: Hyperlocal Citizen Media Sites Want You (to Write)! Online Journalism Review. URL: http://www.ojr.org/ojr/glaser/1098833871.php (accessed 15 January 2012).
Hampton K., Wellman B. 2003. Neighboring in NeTVille: How the Internet Supports Community and Social Capital in a Wired Suburb. City & Community, 2 (4), pp. 277-311.

Lasica J. D. 2005. Citizens' Media Gets Richer. Online Journalism Review. URL: http://www.ojr.org/ojr/stories/090805lasica/ (accessed 15 January 2012).
McLaughlin J. F., Birinyi A. E. 1980. Mapping the Information Business. Cambridge: Harvard University Press.
Mezza M. 2011. Sono le news, bellezza! Vincitori e vinti nella guerra della velocità digitale. Roma: Donzelli.
Monteleone F. 1992. Storia della radio e della televisione in Italia. Un secolo di suoni e di immagini. Venezia: Marsilio.
Noam A. M. 2004. Internet Television: Definition and Prospects. In: A. M. Noam, J. Groebel & D. Gerbarg (eds.). Internet Television. London: Lawrence Erlbaum Associates.
Ortoleva P. 1994. Mediastoria. Mezzi di comunicazione e cambiamento sociale nel mondo contemporaneo. Parma: Pratiche.
Ortoleva P. 2005. La televisione nell'industria culturale, la televisione come industria culturale. In: M. Morcellini (ed.). Il Mediaevo italiano. Industria culturale, TV e tecnologie tra XX e XXI secolo. Roma: Carocci, pp. 273-286.
Pasquinelli M. 2002. Media Activism. Strategie e pratiche della comunicazione indipendente. Roma: Derive Approdi.
Patanè Garsia V. 2004. A guardia di una fede. Gli ultras della Roma siamo noi. Roma: Castelvecchi.
Pescio L. 2007. Storia e prospettive della Web TV. In: F. Colombo (ed.). La digitalizzazione dei media. Roma: Carocci, pp. 87-104.
Richeri G. 2009. I tre cicli della televisione italiana. In: A. Abruzzese & A. Bonomi (eds.). La cultura italiana. Volume IV: Economia e comunicazione. Torino: UTET, pp. 474-493.
Robertson R. 1992. Globalization: Social Theory and Global Culture. London: Sage.
Rosenberry J. 2010. Public Journalism 2.0. In: J. Rosenberry & B. St. John (eds.). Public Journalism 2.0: The Promise and Reality of a Citizen-Engaged Press. New York: Routledge, pp. 133-60.
Sartori L. 2006. Il divario digitale. Internet e le nuove disuguaglianze sociali. Bologna: Il Mulino.
Selwyn N. 2004. Reconsidering Political and Popular Understandings of the Digital Divide. New Media & Society, 6 (3), pp. 341-362.
Tapscott D., Williams A. D. 2006. Wikinomics: How Mass Collaboration Changes Everything. New York: Penguin.
Tinagli, I. 2008. Talento da svendere. Torino: Einaudi.
Wellman B. 2004. The Glocal Village: Internet and Community. The Arts and Science Review, 1 (1), pp. 26-29.
Veltri F. 2005. La rete in movimento: telematica e protesta globale. Soveria Mannelli: Rubbettino.

Part IV

Understanding New Behaviours and Attitudes towards Digital Television

Jakob Bjur

Social Television Ecology – The Misfits and New Viewing Practices

The death of television. We have all heard it being proclaimed before (Gilder, 1994; Katz and Scannel, 2009) but this is far from reality – both in theory and in real life. A more adequate description seems to be that television is subject to broad transformation; in terms of content and services, distributional form, technology and cultural form. This transformation is not limited to television itself but encompasses its audiences, becoming highly fragmentised and accordingly, increasingly polarized (Napoli, 2011; Webster, 2005). Thus, what used to be common, shared and homogenous is today individualized, specialized and heterogeneous (Bjur, 2009). Audiences, as well as societies, have transcended from mass to individualized towards networked (Castells, 1997; Jenkins, 2006). During the last two decades, this line of change has been fuelled by digitalisation and accelerated by an increasingly ubiquitous Internet (Fortunati, 2008; Urry, 2007). Consequently, observers have forecasted that television is about to be revolutionized (Lotz, 2007). Addressed below is *when*, *where*, and *how* this revolution is taking place in everyday viewing situations.

The subject matter of this chapter is to outline the social contours of television viewing. The chapter delivers an unique empirical outline of how traditional television viewing is socially structured, in time and in space, and how patterns of social viewing have transformed during the last decade. Inherent to the present social, cultural and technological shift that is breaking the audience apart, from mass to individualized, is a converging force that brings the audience together by means of networks. A special emphasis of the chapter is to confront evidenced social viewing practices with the next generation of television services and technologies aimed for networked audiences. The common denominator of a spectrum of newborn services and technologies, gathered under the label of Social TV (or Smart TV), is that they adapt to the digital mediascape and try to draw on the condition of “networked individualism” (Wellman et al., 2003) this way “relying on relationships to rebuild TV audiences” (Klym and Montpetit, 2008).

There is no doubt change is underway in the way audiences as well as mediascapes transform, and that the future, in this respect, is already present, to some degree available for scrutiny. Central to the television ecologies of tomorrow is the contemporary transformations of the social television ecology of today. Social interaction, has always been, and still is (as will soon be evidenced) a cornerstone in the foundations of television viewing. Consequently, this chapter delivers empirically informed conclusions regarding potential tensions between well-established practises and new services, together with predictions of when, where and for which parts of the television audience Social TV will eventually work.

The Collapse and Revival of Television

At the beginning of the 2000s, a state of crisis constituted the Zeitgeist of the television industry as whole sectors of media industry seemed to be pushed towards the edge of collapse. Great concerns were raised about the young television audience giving up traditional television viewing, so it was said, in favour of Internet sustained pleasures such as downloading and streaming. The threatening consequence of the emergent Internet sustained practices was that music, film and television content began to circulate beyond established chains of distribution of traditional media institutions, in an uncontrolled fashion. Television was consumed on alternative and unmonitored platforms (first PCs at home, and later paralleled by numerous mobile devices) this way undermining well-established business models of television (Picard, 2002). The perception of this state of affairs at that time, was, that the young were only the first, destined to be followed by a much more comprehensive crowd.

With hindsight the story of Peter and the Wolf comes to mind, as television today seems to be flourishing more than ever. The Super Bowl (the major U.S. television event of the year) has surpassed a viewing record on traditional television established in 1983 for two years in a row. Moreover, the Super Bowl is but one of numerous viewing records, as viewing time figures peak in various television markets worldwide. Though hard to measure, new platforms constitute new streams of revenue that become incorporated in the broadened television business. This was first extended to the Internet and now it is also spreading to the mobile sphere by means of applications for various mobile devices. The present situation seems, in short, to promise television business a healthier future than ever before, and that goes, interestingly enough and surprisingly so, for traditional, so-called "linear" television services.

An alluring suspicion that is beginning to arise among people working in and around the television business is that there is a possibly straightforward causal connection between the new digital media ecology, surrounding television today, and this sudden restoration of television. Network society (Castells, 1997) conditioned by individualization (Beck and Beck-Gernsheim, 2002; Giddens, 1991), media convergence (Jenkins, 2006), mobility and ubiquity (Fortunati, 2008; Urry, 2007) and affluence in content and choice (Webster, 2005) might have turned out to be television's saviour. This seemingly paradoxical development has been attributed to the ability of the digital mediascape to sustain "networked individualism" (Wellman et al., 2003) structured according to new principles following personal interests, preferences and lifestyles. Apprehended in this way, the individualization of television viewing – breaking the earlier established television audience apart and making it increasingly difficult to catch and maintain large audiences (Turow, 2005) – serves simultaneously as a precondition for realigning and re-socialising the television audience, but now layered in a new way (Bjur, 2011). That this is maybe the case could be indicated by facts like the rise of the Internet water-cooler effect (Stelter, 2010). The total number of tweets (on Twitter) tied to the Super Bowl Final 2011 was 4.5 million, during the event. Similar Internet activity can be found related to other sports and entertainment events where the parallel rise of Internet discussions on Facebook, Twitter and diversive specialized forums seem to push mass audiences, driven by individual preferences, interests and lifestyles, into traditional linear television events. The shift has been described as the rise of the phoenix of the networked audiences from the ashes of the mass audience (Klym and Montpetit, 2008).

New Social Television Services

A great number of continuously evolving television and video services and technologies are currently trying to exploit the window of opportunity opened up by the networked audience. Internet has, put bluntly, given rise to three different categories of services. The first are the so-called "on-demand" or "play services" provided by television channels and television service providers. These are mainly alternative distribution services providing live or time-shifted content distribution via the Internet. Within this category, new actors of streaming services, like Netflix, Hulu and Boxee, emerge. Entering the field of television are Google and Apple illustrating that television is presently a fierce battlefield with many diverse players fighting for revenues. The second category is the so-called "Internet water-cooler services" that provide a parallel online communicative universe around television. These services build on social media dynamics and audience members can in Miso, Getclue and Tunerfish, communicate around and during television viewing. Possibilities are sustained to discuss, rate, log into television programmes and to build favourite lists of television content thereby expressing a personal television identity. To some extent these services constitute parallels to location-based services like Govalla, FourSquare or Facebook places, but aiming for the "referential space" of television and entertainment instead of the geographical space of the real world (Bjur, 2009). The third category of services is the set of technological solutions that literally melt together television and social media on the single platform of the television. These services and technologies labelled "Social TV" are developed at MIT Medialab and other universities of technology together with telco and IPTV operators (Klym and Monpetit, 2008).

The importance of the above categorization lies in the provision of a momentary picture of the current impetus of change inherent to television. The sketch aims at a moving target, and the scope is, rather than establishing exact borders between different categories, to establish the field of power wherein television can be revolutionized. The first category represents another distributive dynamic, the second category complementary communicative tools, while the third stands for the merging of social media and television. Over time the borders between categories, already today blurred, will most likely dissolve.

Old Television Viewing Practices

When surveying the actual design of the plethora of new societal television services it seems to be one piece that is to some extent left out of the picture. That piece is the audience and already established patterns of television viewing behaviour. New sets of services are developed today with the aim of exploiting the possibilities opened up by a networked audience. They are built on the assumption that the audience has transcended from mass to individualized and is now fully able to play the role of networked audiences. However, looking back on decades of audience research, or, for that sake, only taking a glimpse around, it is hard to believe that this is really the case. Television is one of our major leisure time activities, and it is, in many cases, undertaken together with family and peers in social situations around the tube where negotiations, chatting, eating and different forms of multi-tasking is a normal code of procedure.

On the most basic level it takes three components to make television viewing take place: a *viewer* watching; programme *content* being watched; and a situational *context* where the viewing is taking place. A social viewer group, that shares the same physical space around the television, will of course to some degree delimit or at least change the scope of services that are built on the assumption of addressing individualized audiences. As audience researcher James Lull once aptly put it when describing the television viewing practice:

> television viewing is constructed by family members; it doesn't just happen. Viewers not only make their own interpretations of shows, they also construct the situations in which viewing takes place and the ways in which acts of viewing, and program content, are put to use at the time of viewing and in the subsequent communications activity (Lull, 1990, p. 148).

At the centre of the empirical study that is now about to follow are these socially constructed *situations* surrounding the sites where television viewing "takes place". These social situations that continuously vary second-to-second are hard to research and study, a reason they can be described as one of the prime residuals of audience research, a factor placed central in theoretical models and left out in empirical research (Bjur, 2007). Provided with a detailed image of these social situation, of when, where and by whom social life is played out around the television, future television and video services could become better designed to blend with and sustain social life.

Mapping Out the Social Television Ecology

What remains to be scrutinized is how these social foundations, inherent to television viewing practices, have changed over time and look like today. Television viewing behaviour has distinguished itself to be a fairly solid pattern of behaviour, strongly aligned to social everyday life and subject to comparatively slow and incremental change (Silverstone, 1994). Is this still the case? Or has the last decade radically changed and dissolved the social foundations that television viewing used to rest upon?

There is no question that mass audiences worldwide are becoming fragmented, but is the micro-level family audience travelling down the same road? Are family audiences even showing tendencies of facing this destiny? One way to find out is to study how television viewing is structured in relation to social everyday life at home, in different households, at different times of the week and day. The present study is a step in this direction aiming to deliver answers to two fundamental research questions regarding social viewing:

RQ1: Where does social viewing occur and *who* is the social viewer? (SPACE)
RQ2: When does social viewing occur? (TIME)

The scope of the remainder of the chapter is to make a fairly detailed delineation of these questions concerning the contemporary social television viewing ecology together with its transformation during the last decade.

Methodology and Data

The presented mapping out of social viewing is based on Swedish People Meter data depicting television viewing behaviour minute-by-minute in a panel of 1,000-1,300 households (2,000-2,600 individuals) selected to represent the national television audience. People Meter technology combines *passive* registration of what the television is tuned into with *active* button-pushing of the individuals registering who is in front of the screen (Webster, Phalen and Lichty, 2000). As with every audience measurement methodology it represents strengths as well as weaknesses (Bogart, 1988; for a methodological in depth discussion see Milavsky, 1992; or Bjur, 2009, Chapter 3). Two unique characteristics of the People

Meter data that make it well adapted for the study at hand is first, its behavioural character – as opposed to self-assessment data, which is the most commonly available but less suitable alternative – and second, the inherent "social design" of the data following measurement of parallel action within households.

In order to map out the social dimension of television viewing, People Meter data has been accessed in raw data form, and then processed and analysed. The methodological principles applied, termed as a procedure of *thickening,* extract socially and culturally meaningful dimensions of viewing behaviour inherent to the data but regularly neglected in the everyday business of professional audience analysis (Bjur, 2011). The necessity of coping with raw data was due to the fact that social viewing is a television viewing behaviour that is not readily available in the software packages for professional audience analysis. The analysis is for each time point (1999, 2002, 2005 and 2008) based on accumulated individual level viewing behaviour of eight consecutive months (September to April). While the years were chosen to account for continuous over time development 1999-2008, the sample of months chosen was based on a criteria of quality – People Meter data is the most robust when television viewing is the most extensive and the audience is based at home (in tab rates drop significantly during summer holidays).

The obvious limitation of the People Meter data is that it is capable of gathering information on traditional television viewing taking place within the household. Consequently the data is indicative of only the most conservative form of television viewing (mainly schedule based home viewing), and the patterns found will, in the concluding part, be discussed from the perspective of what they mean in relation to new forms of television viewing increasingly freed from time (temporal schedules) as well as from space (physical household settings).

Every empirical investigation has to be staged somewhere, and in this case the stage is Sweden. When it comes to television, Sweden is a case that is both particular and general. It is particular, as a strong public service environment deregulated comparatively late, around 1990. The average television viewing time per day in Sweden has been rising from approximately 146 minutes/day in 1999 to 166 minutes/day in 2009. Meanwhile, the daily reach has dropped from 76 to 70 percent (MMS, 2000, 2010). The average availability of channels was 5 in 1999 and 17 in 2009 (Carlsson and Facht, 2010). Swedes are thus modest television viewers in international comparison. On the other hand, when aiming for the future, Internet penetration and use is far developed, and Sweden was in 2010, among the top ranked in three different well-established international indexes depicting Internet maturity (Findahl, 2010).

Arguably, Sweden is as good a case study as any other national case, when it comes to general trends of development transforming television systems worldwide. There is no reason to believe the results below would be radically divergent, in terms of general trends, from other national television context. Actual levels of social viewing might of course differ from nation to nation, but the general patterns found, regarding when and where social viewing occurs, as well as who is the social viewer, do most probably own a more general representability for internationally widespread patterns of social television viewing.

Where Social Viewing Takes Place (SPACE I)

A most seminal spatial factor affecting where television viewing takes place is the size of the household in terms of its members. On the most basic descriptive level, the composition of the viewer group is a number of persons: monads, dyads, triads, quartets, quintets etc. The number of persons viewing television together constitutes the social situation within which choices of content are negotiated and made. This social situation will also intersect with all parts of the reception process affecting perception, attention, interpretation, emotion and satisfaction of the viewers. Constellations constituted of more than one person are for natural reasons more common in multi-person households than in single person households. Figures 1a and b present the composition of the television audience in terms of size of viewer groups.

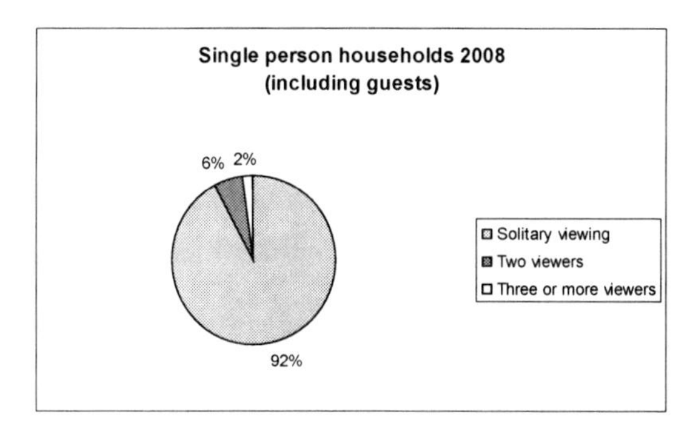

Figure 1a. The size of the viewer group in all viewing 2008 in single households – all viewing (percent of viewing time). N: 2,439.

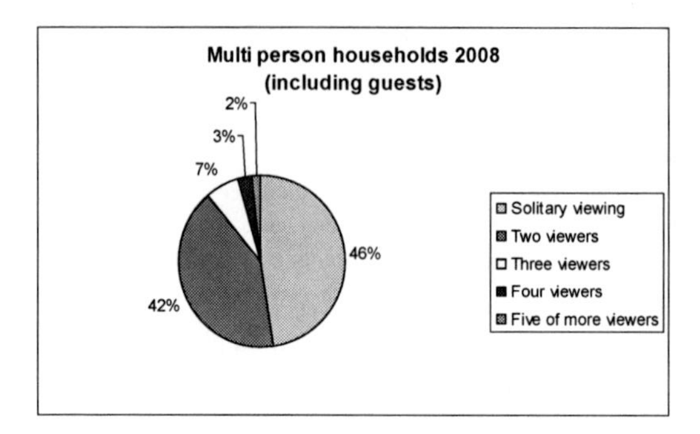

Figure 1b. The size of the viewer group in all viewing 2008 in multi-person households – all viewing (percent of viewing time). N: 2,439.

The dyad is the most usual form of social viewing. In a multi-person household, the situation of two persons viewing together occurs almost as often as solitary viewing – 42 against 46 percent respectively. Television viewing in larger constellations, such as triads, quartets, and quintets and larger is less frequent and falls in amount with increased size. In multi-person households triads account for seven percent of the television viewing, quartets three percent, and quintets and larger two percent. In single person households, the level of social viewing is significantly lower (eight percent). Around four-fifths (78 percent) of the social viewing – or six percent of the total viewing in single person households – is undertaken in dyads. A more detailed picture of the size of the viewer group can be assessed in Figures 2a and b dividing resident household member viewing and guest viewing.

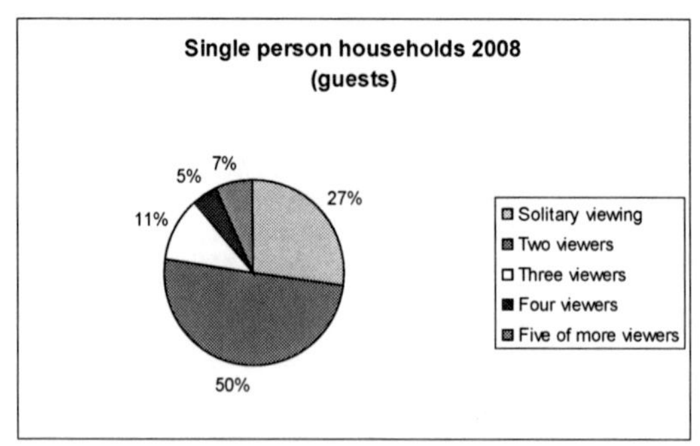

Figure 2a. The size of the viewer group in guest viewing 2008 in single households – guest viewing (percent of viewing time). N: 2,439.

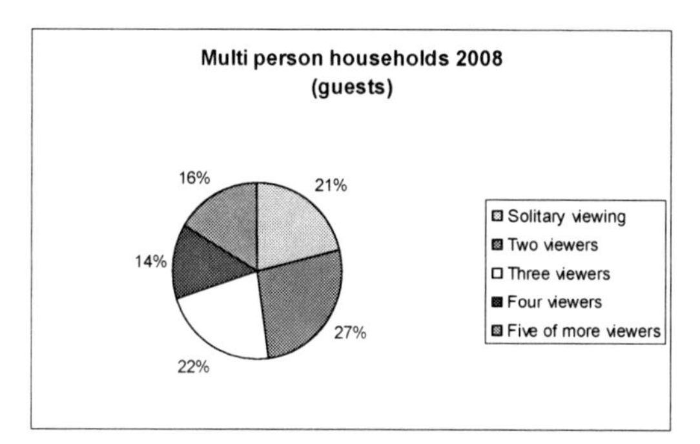

Figure 2b. The size of the viewer group in guest viewing 2008 in multi-person households – guest viewing (percent of viewing time). N: 2,439.

Guest viewing is the television viewing performed by visitors from outside of the household such as visiting friends and relatives. Guest viewing diverges from other categories of viewing when it comes to size of the viewer group. The natural reason for this is that the act of watching television at someone else's home is different compared to the practiced television viewing at home. Solitary viewing is less common and encompasses only around one-fifth of the viewing in multi-person households and one-quarter of the viewing in single person households. The amount of viewing undertaken alone, in dyads, in triads, in quartets, and in quintets and larger are fairly evenly distributed for guest viewing. This is especially evident in multi-person households but also in single person households. The dyad is still the most frequent viewer group size but constellations with more viewers than two are much more frequent in guest viewing compared to resident household member viewing.

Guest viewing is a highly socialized type of television viewing. The over time change seems to imply a slow loosening up of this social dimension of guest viewing as people visiting each other increasingly consume television alone. To watch television alone at someone else's home is still in 2008, a minor category of guest viewing, but a minor category of viewing that is growing over time. In 1999, only 20 percent of the guest viewing was done alone in single person households. The share in multi-person households was 16 percent. In 2008, the corresponding levels of solitary guest viewing have been raised to 27 respectively 21 percent of viewing time. From a perspective of individualization, it is an interesting trend that the act of visiting one another, by definition a social act, is increasingly filled with a solitary practice.

However, the relatively small size of guest viewing (five percent of viewing time) makes it close to insignificant for the general picture. Dominating is the impact of home viewing – representing 95 percent of the total viewing – which makes the total distribution of numeric viewer constellations, in Figure 1 above, differ only marginally from the picture of household member viewing given in Figure 3 below.

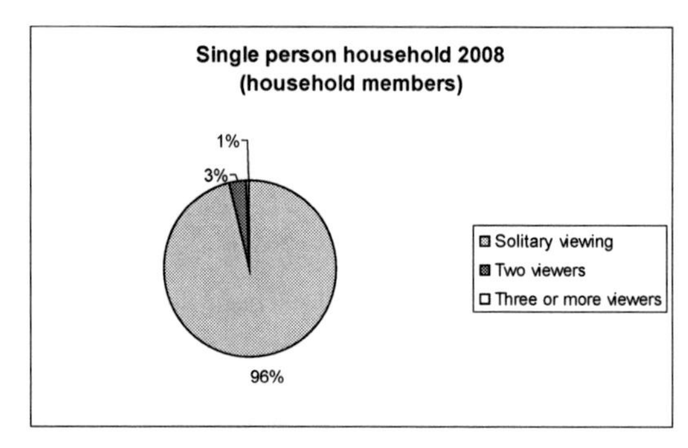

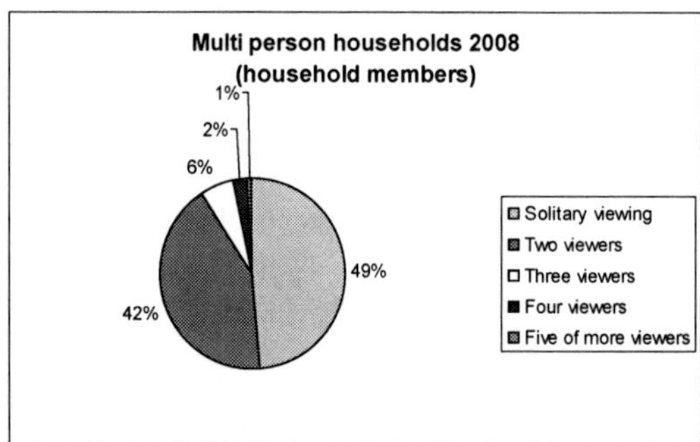

Figure 3. The size of the viewer group in resident household member viewing 2008 in single and multi-person households – household member viewing (percent of viewing time). N: 2,439.

A number of conclusions can be drawn from the display of the numeric viewer constellations of the television audience displayed above. The overarching conclusion is: depending on which type of household – single person or multi-person – and which type of television viewing you focus on – viewing at home or guest viewing – the expectation of social interaction in television viewing differ. The social setting of the household has a clear baseline effect for social patterns of television viewing that becomes evident when you split single person and multi-person households. Television viewing in single person households is by default individual and consequently most often solitary. The inherently asocial characteristic of single person households means that for television viewing to become social, it has to involve guests from outside the household – a situation occurring four percent of the time of viewing.

Likewise, guest viewing can be said to be inherently social. Seldom do guests enter someone else's home to watch television alone. In single-person households, this situation occurs one-quarter of the viewing time and in multi-person households one-fifth of the viewing time. This is low compared to the level of solitary viewing of resident household members when at home – 96 and 49 percent respectively. The conclusion that can be drawn is that the act of visiting a friend – that arguably is interaction between households – more often results in social than solitary patterns of consumption when it comes to the practice of television viewing. A conclusion that is more comforting from a humanitarian perspective than surprising from a commonsense perspective. But as has been found, it has been contested by development over time.

The conclusions presented above have consequences for the way social viewing is to be researched. First, as stated previously, single person households can supply no information when it comes to questions on intra-household interaction, since they represent none (following the simple law that it takes two to tango). Second, when guest viewing is included in analysis of social viewing its inherently social character will bias results in the direction of indicating more social behaviour than present among the actual members of the household. Consequently, both these categories of viewing – viewing within the single person household setting and guest viewing – have to be distinguished when answering the question of who the social viewer is.

The Composition of the Social Audience (SPACE II)

Before going into detail concerning additional characteristics of the social viewer it must be underlined that the split between single and multi-person households is seminal. Even if other characteristics such as age, number of television sets and available channels and zapping behaviour are significant factors with ties to social viewing behaviours, the split between asocial and social ground is of another magnitude. So the reader should keep in mind social viewing takes place in multi-person households, and is of only marginal occurrence in single person households.

However, three more specific conclusions about the social dynamics of television viewing can be drawn from a more profound study on the patterns of social viewing of individuals resident in multi-person households. The first one concerns the reasonable expectation that availability of television equipment and the amount of channels would have an individualizing effect on television viewing behaviour. Social television viewing turns out to be a less comprehensive phenomenon in multi-person households with more than one television set and for individuals using a broader spectrum of channels. The composed effect of these two technology factors is slightly more influential than the effect of age or the number of persons in the household. Excluding the youngest, social viewing increases linearly with age. The youngest and the oldest share the highest levels of social viewing. Levels of social viewing increase with the number of resident household members.

The second conclusion regards how viewing behaviours play a role in guiding patterns of social interaction around the television. Individuals with a stronger mobility over the channel flows (zappers) show a higher degree of individualization in their viewing. This finding could be the other way around: when viewing together with others extensive zapping is less common.

A third conclusion, of no less importance than the first two, relates to all those factors that did not show any or weak interrelations with social viewing. Weak or unstable interrelations, where represented by gender, education, availability of video, PC and Internet, television viewing together with video and gaming time – as viewing behaviours. An intermediate position of fairly strong connections to social viewing are held by form of dwelling – influential to a varying degree – and by the way of receiving the signal – that lost its significance as all Swedish television households as a consequence of digitalisation of the terrestrial network 2005-2007 were turned into multi-channel settings.

One of the most interesting overall findings of the study was that all the measured factors lost explanatory power over time. This means today it is much harder, than a decade ago, to predict who is to perform a social and who solitary (or individualised) viewing behaviour. The television audience is becoming increasingly hard to predict based on individual characteristics, such as demographics, and structural factors, such as technique availability and social setting. The television audience is increasingly individualized.

When Social Viewing Takes Place (TIME)

Social patterns of television viewing exhibit a strong link between everyday life and the practice of television viewing. During the working week, leisure time is limited to morning time and evening time for a majority of the audience. Weekends constitute a break from this nine-to-five working week, opening up a social leisure time where television viewing can be performed, as one among many leisure time practices. Consequently, to come closer to real world television viewing behaviour, it is necessary to split viewing patterns over weekdays and weekends. This way it is possible to see if change is especially tied to a particular section of the week, i.e. weekday or weekend, and these working week patterns of television viewing is illustrated in the graphs below where rating curves are used to outline how television viewing is distributed over the day, weekdays (Figure 4) and weekends (Figure 5).

The total rating shows the size of the audience that tune in at different times of the day while the social rating tells the proportion of the audience watching television in a social situation together with other viewers. The brighter top field and the upper dotted curve of each graph show the shift in overall viewing, while the darker bottom field and lower integer curve outline the parallel development of social viewing. In both graphs, the two grey areas of 1999 respectively the two black curves of 2008 are laid out fairly close to each other illustrating the incremental over time change in television viewing behaviour during one decade.

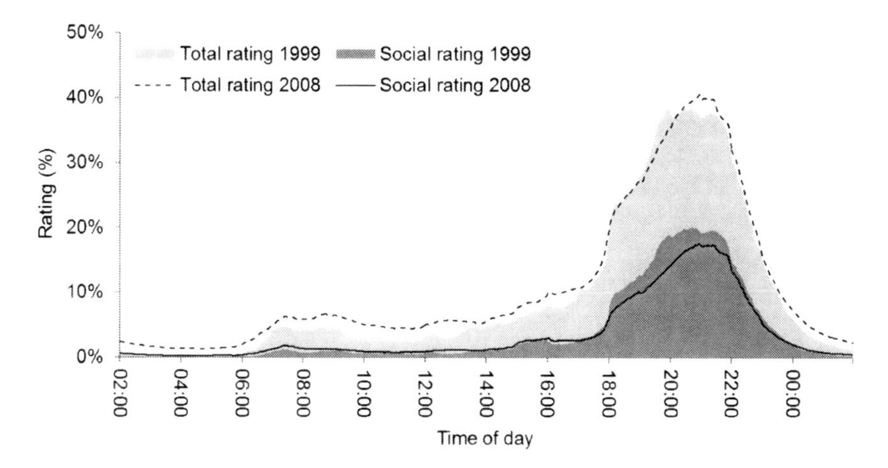

Figure 4. Rating and social rating curves – an average weekday (Monday to Thursday) 1999 and 2008 (percent of audience). N: (1999)=2,439 (2008)=3,137.

Figure 4 illustrates that social ratings follow overall ratings most times of the day during weekdays and weekends. Daytime (06:00 to 18:00) and night-time (midnight to 06:00) both total and social ratings are more comprehensive in 2008 than in 1999 following increased television viewing spread over the day. The more interesting divergence between years occur in broader prime time (18:00 to 23:00) where the largest proportion of television viewing time is invested and the audience is most highly valued by broadcasters and advertisers. The shift in overall prime time viewing consists in a *compression*, making the peak in rating reach higher weekdays, and a *dislocation* of prime time viewing until later in the evening, at weekends. Social viewing follows this trend of compression and dislocation, the difference being social viewing simultaneously shrink. This is shown by the lower black curves of social rating (2008) being positioned upon the lower dark grey field of social viewing (1999). This situation applies 18:00 to 23:30 weekdays and 18:00 to 22:00 weekends. Consequently prime time is the delimited time slot where the social audience is dissolving over time while the social audience increases in amount at all other times of the day and night.

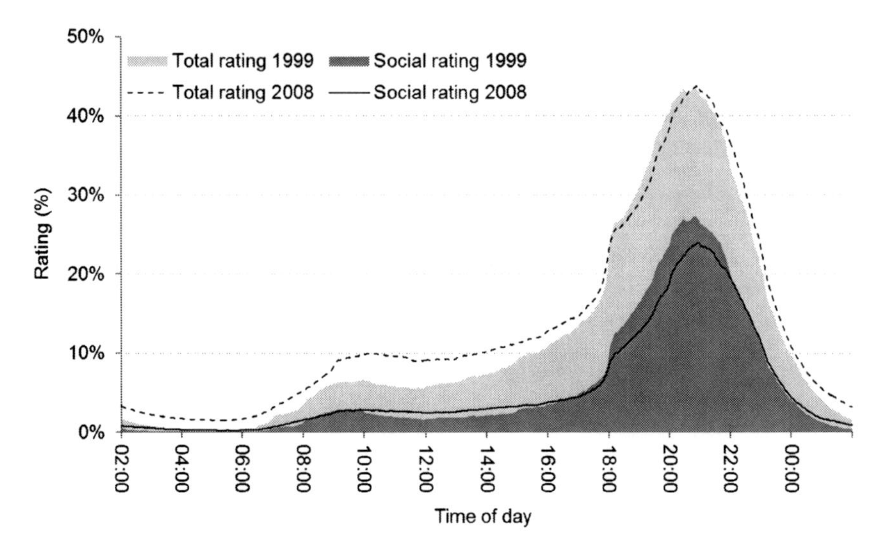

Figure 5. Rating and social rating curves – an average weekend (Friday to Sunday) 1999 and 2008 (percent of audience). N: (1999)=2,439 (2008)=3,137.

That the social audience is dissolving at prime time means that the largest volumes of television viewing today are observed more individually than they used to be. A prime time audience composed of families is, over time, becoming a less suitable guideline for contemporary scheduling strategies as singular family members increasingly consume television individually. These changes in social viewing are concrete changes in physical patterns of interaction around the television located in the physical setting of the home where television viewing takes place. That television viewing is losing social ground is something that opens up an increased space for individual consumption at the intimate site where television viewing comes about. These changes in *physical space* will imply changes in what family members share in terms of consumed content mediated by television, which is a question that can and has been studied in terms of shared *referential space* (Bjur, 2009).

In order to get a balanced perspective, it is important to underline that even if the decline of social viewing and family viewing is broad, there are still social settings where television viewing, to a large extent, equals social television viewing. For 2008, in multi-person households at the peak of prime time (21:00), the average share of social viewing is 66 percent of the total viewing. Consequently, it is important to acknowledge that television viewing is still a far from individualized activity in multi-person households at the time when viewing peaks. This fact has to be taken account of when developing future television and video services for a networked audience.

Social Television Meets Social Television Practices

To conclude, we can now return to the audience transcending from mass to individualized to networked. As can be seen from the results presented in this chapter, the process of individualization is in progress, but far from concluded. The levels of social viewing in single and multi-person households (different spaces) show respectively the distribution of social viewing over the week and day (different times) and that the social viewing still constitutes a kind of ruling condition in certain spatial contexts at certain hours (such as the multi-person households during prime time at weekends). Meanwhile, other spatial contexts, such as the single person household, are well aligned with the description of fully individualized and networked audiences.

The split of television viewing into solitary and social viewing provides analysis of television viewing with a social leverage (Bjur, 2011). This social leverage lends some guidelines for the appropriation of future television services aiming to blend television with social media dynamics. The first conclusion that can be drawn from the presented results is that single-person households constitute a segment of the audience that is well suited to adapting services allowing the television experience to be socially shared with distant others. Young people living by themselves at a distance from relatives and peers could, for example, share television viewing situations and experiences, a case that could also apply for senior relatives, the older mother or father living separated from children and grandchildren. These groups could be rewarded achieving co-presence at a distance, and in these specific situational circumstances the described services could actually sustain and fuel social cohesion.

The more problematic side of the coin is how television juxtaposed with social media is to be adapted to the highly social situations of television viewing presently dominating multi-person households at prime time. In the U.S., where television has been identified with undermining the social capital of American civil society (Putnam, 2000), Social TV services are seldom packaged as technologies that will finally change this condition, turning television into a net contributor of social cohesion. Faced with the empirical reality laid out above it is hard to see exactly how this is going to be the general case. Social television promises in itself extremely interesting future features, as the black boxes tied to our television set get connected to the black boxes of distant others and we also start sharing and distributing mediated experiences within our personal networks. However, it is hard to see how Social TV can set back the continuous breakdown of the micro-level family audience.

When gathered around the television, every single individual is able to individually make use of the Internet water-cooler and to react and communicate with distant others, at the same time being physically co-present with partners and family. That being said, when it comes to the choice of what is presently running on the shared screen, however, the opportunities for viewers who are not co-present are most likely reduced. Part of the individualization of television viewing can be attributed to the fact that television content over time has been increasingly segmented to suit different well-defined market segments, divergent in age and gender. This is a development that is rather poorly suited for a family audience when designated channels are tailored to appeal to different family members (Ettema and Whitney, 1994).

To conclude we can remind ourselves of the strategy of the mother Babette, in Don DeLillo's *White Noise*. On Friday nights, Babette made it a rule for the whole family to watch together while eating take-out Chinese food. She believed:

> the effect would be to de-glamorize the medium in their eyes, make it wholesome, domestic sport. Its narcotic undertow and eerie diseased brain-sucking power would be gradually reduced (DeLillo, 1986, p. 16).

Babette's strategy was, as far as readers could tell from the novel, of limited success. It remains to be seen if future social television projects manage better.

References

Beck U., Beck-Gernsheim E. 2002. Individualization – Institutionalized Individualism and its Social and Political Consequences. London: Sage.

Bjur J. 2007. Social Television Viewing: A Reassessment of the Prime Time Residual of Audience Measurement. Paper presented at the Applied Econometrics Association (AEA) Conference MEDIA and COMMUNICATION Quantitative and Econometric Analysis. Paris, Sorbonne, 22-23 November.

Bjur J. 2009. Transforming Audiences. Patterns of Individualization in Television Viewing. PhD dissertation, University of Gothenburg. URL: http://gupea.ub.gu.se/handle/2077/21544?mode=full (accessed 18 January 2012).

Bjur J. 2011. Thickening Behavioural Data: Toward an Increased Meaningfulness in Behavioural Data. Conference on New Challenges and Methodological Innovations in European Media Audience Research. COST-IS0906, Transforming Audiences. Transforming Societies. Zagreb, 7-9 April.

Bogart L. 1988. Research as an Instrument of Power. Gannett Center Journal: Measuring the Audience, 2 (3), pp. 2-17.
Carlsson U., Facht U. (eds.). 2010. MedieSverige 2010 – Statistik och analys. Nordicom, University of Gothenburg. Gothenburg: Nordicom.
Castells M. 1997. The Information Age. Economy, Society and Culture. Volume I: The Rise of Network Society (2nd ed.). Oxford: Blackwell Publishers Ltd.
DeLillo D. 1986. White Noise. New York: Penguin Books.
Ettema J. S., Whitney C. D. 1994. Audiencemaking: How the Media Create the Audience. Sage Annual Reviews of Communication Research, 22, pp. 1-18.
Findahl O. 2010. Svenskarna och Internet. Gävle: World Internet Institute.
Fortunati L. 2008. Mobile Convergence. In K. Nyiri (ed.). Integration and Ubiquity. Towards a Philosophy of Telecommunications Convergence. Vienna: Passagen Verlag, pp. 221-228.
Giddens A. 1991. Modernity and Self-identity: Self and Society in the Late Modern Age. Cambridge: Polity Press.
Gilder G. 1994. Life After Television. New York: W.W. Norton.
Jenkins H. 2006. Convergence Culture – Where Old and New Media Collide. New York: New York University Press.
Katz E., Scannel P. (eds.). 2009. The End of Television? Its Impact of the World (So Far). The Annals of the American Academy of Political and Social Science, 625, Thousand Oaks, CA: Sage. pp. 6-18.
Klym N., Montpetit M. J. 2008. Innovation at the Edge: Social TV and Beyond. MIT CFP – VCDWG Working Papers. Cambridge: MIT.
Lotz A. D. 2007. The Television Will Be Revolutionized. New York: New York University Press.
Lull J. 1990. Inside Family Viewing – Ethnographic Research on Television's Audiences. London: Routledge.
Milavsky J. R. 1992. How Good is the A. C. Nielsen People-Meter System? A Review of the Report by the Committee on Nationwide Television Audience Measurement. Public Opinion Quarterly, 56 (1), pp. 102-115.
MMS 2011. Årsrapport 2010. Stockholm: MMS.
Napoli P. M. 2011. Audience Evolution: New Technologies and the Transformation of Media Audiences. New York: Columbia University Press.
Picard R. G. 2002. The Economics and Financing of Media Companies. New York: Fordham University Press.
Putnam R. D. 2000. Bowling Alone: The Collapse and Revival of American Community. New York: Simon & Schuster.
Silverstone R. 1994. Television and Everyday Life. London: Routledge.
Stelter B. 2010. Water-Cooler Effect: Internet Can Be TV's Friend. New York Times, 23 February 23. p. A1 (New York Edition).

Turow J. 2005. Audience Construction and Culture Production: Marketing Surveillance in the Digital Age. The Annals of the American Academy of Political and Social Science 597 (1), pp. 103-121.
Urry J. 2007. Mobilities. Cambridge: Polity Press.
Webster J. G. 2005. Beneath the Veneer of Fragmentation: Television Audience Polarization in a Multichannel World. Journal of Communication, 55 (2), pp. 366-382.
Webster J. G., Phalen P. F., Lichty L.W. 2000. Ratings Analysis – The Theory and Practice of Audience Research (2nd ed.). Hillsdale, NJ: Lawrence Erlbaum Associates, Inc.
Wellman B., Quan-Haase A., Boase J., Chen W., Hampton K., Díaz I., Miyata, K. 2003. The Social Affordances of the Internet for Networked Individualism. Journal of Computer-Mediated Communication, 8.

Bartolomeo Sapio, Tomaz Turk, Stefano Livi, Michele Cornacchi, Enrico Nicolò & Filomena Papa

User Experience of Payment Services through Digital Television

Introduction

Developing new technology which is acceptable to humans is a complex process, especially when the setting of the operational criteria for efficiency, effectiveness and end-users' satisfaction has to be considered as the central activity in the implementation. An example of this is the changing television ecosystem which now offers great potentials to investigate evolving user experiences with new interactive services.

The main objective of this paper is thus to identify a methodology, centred on knowledge about user experience and user acceptance, to investigate diffusion patterns of payment services through DTV.

The chapter examines a microsimulation diffusion model for payment services through digital television (DTV) as it is applied to an Italian field study in order to better understand the impact on citizens, their behaviours during the early adoption phase and the main conditions for optimal use. The final goal is to predict the implications of possible strategic measures, thereby foreseeing the effects of actions carried out by public and private stakeholders in different operational contexts.

Accordingly the chapter presents a composite user-driven reference method of usage behaviour forecasting. When a new emerging technology is somehow proposed to citizens several factors may concur to determine the general intention of use. This study examines the relevant predictors of digital television (DTV), as basically set out by the UTAUT (Unified Theory of Acceptance and Use of Technology) model, in order to understand which of them are the most significant to the intention of use. Those variables are later processed by a microsimulation model in order to identify a representation trend of the final adopters. This is also dependent on the governmental policies applied to sustain the penetration of DTV. The study, although based on a limited sample, nevertheless points out the huge influence of end-user variables in determining the success or the partial failure of the adoption strategy. In particular, the perception of security has been recognized as a crucial variable when a new interactive payment service is proposed.

The next section sets out the approach adopted in this chapter to explore and examine the user experience.

A User-Driven Reference Method

As mentioned earlier, DTV adoption behaviour is explored here by utilizing the UTAUT model that provides an integrated view of user acceptance. This model has been positively applied in several fields when emerging ICTs are implied, e.g. in order to understand the adoption determinants of Internet banking to improve services and attract more users (Foon and Fah, 2011), to estimate factors affecting usage intention of mobile commerce (Tao Zhou, 2008); to determine the extent to which students accept the mobile-learning education delivery methodology (Williams, 2009); to enhance understanding of the SME adoption of wireless technologies (Anderson and Schwager, 2003); to explore adoption of ICT in a government organization and enhance government-to-employee interactions (Gupta, Dasgupta and Gupta, 2008), and to understand tendencies of students determined by age and gender in using course management software (Marchewka, Liu and Kostiwa, 2007).

The convergence of the ICT and the broadcasting world is faced with the uncertainty and the slowness of the users in adopting new opportunities. As a result of many classes of users may not be equipped with the necessary cognitive skills to catch all the potential advantages offered by the technology evolution. In particular, the relationship between users and digital technologies is often ambivalent if we consider, from one side, the possibility of accessing massive informative resources and services, and from the other, the growing complexity of the human interaction with new innovative digital devices. There are clear upper boundaries to which the users are urged to go, for example those of the usage model, terminology, graphical interface, style of use, and so on, where the essential needs and expectations of the end-users are often placed in a second layer and interaction routes are rigid and predefined. There is no comprehensive observance of the fact that ICT innovation may vary notably from the perspective of the end-users reaction to the change. These changes are occurring at great speed and may give rise to two types of reactions: users quickly get used to changes, benefit from the advantages and make them a natural part of their lives; alternatively, users who have difficulty in assimilating technological improvements into their own everyday lives, keep lagging behind.

The governmental bodies are responsible for rethinking the system/services by focusing on the citizens for extensive provision and digital innovation (OECD, 2009). They approach this significant problem of the digital divide by setting up explicit policies to sustain and facilitate those who are easily joining the digital society and those who are not (Monti, 2010).

The introduction of the digital terrestrial television has not always adequately taken into account the behavioural model of the pre-existing customer population. In particular, it does not seem to have included specific usability studies at the stages of the development in order to avoid future rejection or usage under expectations. It is nonetheless matter of fact that unsatisfactory early experiences in interacting with new service/system may produce not acceptance but frustration and premature abandonment of the so-called innovation. Furthermore, an incorrect approach may produce distorted ideas about newly proposed digital opportunities, and reduce the perceived sense of security the end-users should have, for example, in approaching interactive services implying money transactions.

This chapter focuses on the important novelty introduced in Italy by the Digital Terrestrial Television (DTV introduction currently in progress) with respect to the old analogue system. DTV brings a variety of new options with which the citizens can effectively modify their own traditional passive role in front of the television and acquire an active style of use (Cornacchia et al., 2008a). Digital Television is undoubtedly intended as a new opportunity for people to be emotively involved into the general underlying process of cultural change in society by lowering the digital divide. This appears in part as a user-driven innovation, meaning that responsible parties listen to customers or key actors in the early phases of the adoption process of a new interactive service, or later, as feedback is generated.

The emphasis of this chapter is to look at the public service process of acceptance from the end-user perspective (Cornacchia et al., 2008b) and identify those significant elements which cut across the service (T-bollettino) giving the idea of the future adoption trends. The ultimate purpose is to analyse whether this exercise is viable and if it is possible to obtain a reference method, applicable to other similar emerging technologies placing the user in the centre for services involving a composition of building blocks. End-users in this case are the greatest contributors to the innovation process, such as when they are given the possibility of participating in the creation of new services they could properly use, and that they could mould to satisfy their immediate and fundamental needs – and that they could enjoy.

In order to clearly understand the behavioural changes occurring with DTV in Italy (Livi, et al., 2010), we took advantage of a different multidisciplinary approach by outlining a method from different perspectives. For this purpose, we employed the case study of T-bollettino, an interactive payment service associated to DTV, in order to analyze end-user behaviours occurring within the selected event.

The case study is a scenario where we can explore the processes occurring in people's lives and, from these, point out their main needs and concerns about the T-bollettino. From this analysis we have extracted elements which help to understand the outcomes of a predictive model (microsimulation) of adoption for the service. The main objective, then, is not an exhaustive review of the service, but a feasibility check of a user-driven reference method, shaped for interactive services and intended as an opportunity on social growth through DTV development.

Finally, in striving to provide an effective experience to end-users as close to being involved as possible, the two majors issues considered, namely usability

(Performance Expectancy) and security (Perceived Security), have shown both a central influence on the method proposed and a deep correlation (Papa et al., 2010). For example, without extensive customisation users are often presented with many options (Yee, 2002) that security policies may prevent them from using or alternatively is mandated, thus introducing either confusion or frustration (Kluever and Zanibbi, 2008). In order to have a successful new DTV service, it must be understood by novices and inspire confidence, especially if money transactions are proposed.

The T-Government Project "Services for citizens via DTV"

In this section the main features of the field investigation developed in the Italian T-government project "Services for citizens via DTV" are summarized (Papa et al, 2010).

Objective

The main goal of this project was to investigate the user experience and the user behaviour in payment services realised by DTV based on modern theories on the Technology Acceptance Model (i.e., Venkatesh and Davis, 2000). The Classic Technology Acceptance Model (TAM) suggests that when users are presented with technological devices, a number of factors influence their decision about how and when they will use them, notably two main key sets of constructs: Perceived Usefulness and Perceived Ease of Use (Davis, 1989). UTAUT extends TAM by introducing the terms of social influence and cognitive instrumental processes and, above all, it ultimately unifies the main competing user acceptance models, namely eight theoretical approaches sharing the same basic concepts. Actually the TAM and UTAUT (Unified Theory of Acceptance and Use of Technology) were both applied to the adoption and use processes of emerging ICTs in order to understand the human choices and technology acceptance (Livi et al. 2010, Sapio et al., 2010). Thus, the main objective was to evaluate the influence of usability and economic aspects on the user's decision to adopt and to use a DTV service for payments based on the UTAUT model, using a microsimulation approach.

Approach

The field investigation with real users was realised by referring to human factors guidelines. The following main user experience aspects related to interactive services were investigated according to the UTAUT model of technology acceptance (Venkatesh et al., 2003): perceived usefulness, perceived ease of use and attractiveness, training and user support (human support, user manual, support provided by DTV, call center), user perception of technical disturbances and troubles (due to television signal, set top box, return channel), security and privacy perception (confidentiality of personal data, security of payments), impact of the equipment in the house, users' satisfaction about the service including the comparison of different channels to perform the same task (e.g. DTV versus Internet, DTV versus traditional office desk).

The "T-Bollettino" Service

T-bollettino is an interactive service providing the user with:

- remote interactivity using the return channel
- user identification and authentication using a smart card
- on line payment functionality

The return channel is connected to the DTV service centre of the project, and the broadcasting system broadcasts the application. The service allows bills payment through DTV for gas, phone, local taxes, fines and road taxes. The payment can be made using a credit card, a prepaid card (Postepay) or by charging the account "Bancoposta Online".

Users' Panel

A panel of about 300 users was involved in the field study. The panel was uniformly distributed in the North, Centre (City of Roma) and South of Italy. The users were selected with the criteria to be representative of Italian families in terms of number of components of the family, age of head of the family, education of head of the family, job of head of the family.

Only people usually paying bills at the post office and owners of a credit card, a prepaid card (Postepay) or the current account "Bancoposta" were included in the users' panel.

Procedure

In the first phase of the project, the T-bollettino service was developed and broadcast in the whole country. In the second phase the service was experimented in the field with real users. A decoder self-installation procedure was adopted supplying the user with an adequate user guide. During the field study a call centre was available to the users giving any kind of information and help for decoder installation and for service utilisation.

Tools and Techniques for Data Collection

The data collection was mainly carried out by administering to the users a semi-structured questionnaire using a CATI (Computer Aided Telephone Interview) technique.

Regression Model

A dataset referring to 189 subjects was considered in the analysis. Among the 300 users initially selected, only 189 users answered the question about the behavioral intention to use T-bollettino. This lower number of respondents can be related to different causes: some users did not receive the package containing the set top box, some users did not complete the self-installation procedure, others were not able to receive the DTV signal. In the end, 100 users effectively tried to pay a bill through T-bollettino.

In order to better understand the active relations among the newly generated set of variables, a regression analysis was applied in accordance with the UTAUT reference model and methodological literature (Venkatesh et al, 2003). This choice was made in order to test the UTAUT predictive model and to select the significant variables, in order to submit to the microsimulation based estimate the different weights of the UTAUT model factors. The model was thus tested using as criteria behavioural intentions to use DTV payments in future. As shown in Table 1, all coefficients with the exception of Effort Expectancy were significant (overall model: Adjusted R Square=.15; p<.000). Hence, it can be said, as system performance (beta=.24), presence of facilitating conditions (beta=.15) and perceived secure payment system (beta=.25) increase, it encourages the intentions to use the DTV payment services in future.

Predictors	B	S.E.	Beta	t	Sig.
Constant	1.427	1.121		1.273	0.205
Performance	0.296	0.082	0.243	3.586	0.001
Effort	-0.047	0.143	-0.022	-0.330	0.742
Facilitating conditions	0.199	0.093	0.146	2.143	0.033
Perceived security	0.231	0.064	0.251	3.588	0.001

Table 1. Regression model predicting behavioural intentions to use the DTV payment services in future.

Surprisingly, effort expectancy was not a significant variable in predicting intentions with a coefficient close to zero. For this reason this variable was removed from subsequent microsimulation analysis. Finally, perceived security was found to be moderately correlated with other UTAUT predictors, while experienced effort, performance and facilitating condition was uncorrelated as predicted by the theoretical model (see Figure 1).

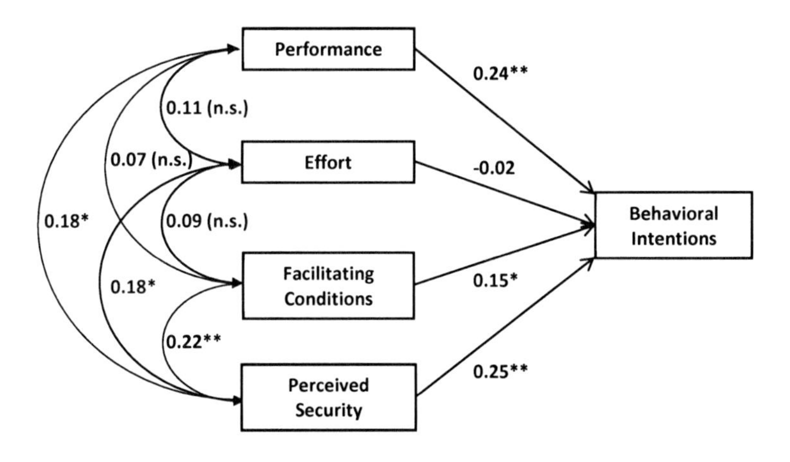

Figure 1. Standardized regression coefficients predicting behavioural intentions to use the DTV payment services in future.

Microsimulation Model

The purpose of our microsimulation model, especially designed for this study, is to enable the exploration of the effects of different policies upon the usage behaviour of adopters. The model was developed in such a way that the usage behaviour is shown through the number of actual interactive service users, which could vary according to different scenarios ("policy packages"). We used the microsimulations approach since it takes into account each person individually. Changes in policies influence each person in such a way that her decisions could change (e.g. decision to use an interactive service).

Since the outcome variable is the number of interactive services users, one of the assumptions is that a person is already using DTV as a platform, so the simulation model should firstly predict the number of DTV users. To do this, our solution incorporates the Bass diffusion model (1969) on the aggregate level. From this the number of DTV users is derived (dynamically in simulated time). The simulation is developed in such a way that firstly the number of DTV adopters in next time period (year) is calculated, and then the actual persons in micropopulation are randomly selected as DTV adopters. This introduces certain variability in the results.

As mentioned, our solution incorporates the Bass diffusion model for DTV diffusion, and the UTAUT model for interactive services adoption. The reason for this approach was the fact that we had no data on the individual level from which we could estimate the user behaviour regarding the DTV adoption process. The Bass model seems a reasonable choice, since it is an aggregate model which shows the dynamics of diffusion of technology.

An important point is to distinguish between diffusion and adoption of technology. In the analysis of adoption one considers the decisions taken by agents to incorporate a new technology in their activities. By contrast, in the analysis of diffusion one is concerned with measuring the change of economic significance of a technology with the passage of time. In a sense, the analysis of diffusion is closely related to the analysis of technological substitution in which the displacement of one technology by another is the focus of attention. The spread of new technology occurs in a number of dimensions (Korres, 2008).

Diffusion patterns can be studied in different ways. In our analysis, aggregate modelling was chosen since it is argued that an innovation's likely diffusion pattern within a country can be better analyzed with an aggregate, Bass-type diffusion model (Bass, 1969) than with disaggregated data. Our study mainly focuses on the within-country diffusion process, i.e. given the time of adoption what is the innovation's likely diffusion pattern within a country (Dekimpe et. al., 2000).

The most popular models of this type are the Bass model and the Gompertz curve (for a discussion, see Meade and Islam, 1995). Both models are relatively simple in structure and provide a set of parameters which express the nature of the studied diffusion process. The Bass model gives estimates for the total number of potential adopters, the coefficient of innovation and the coefficient of imitation. The Gompertz model is usually considered as an "internal influence" model where later adopters learn from earlier adopters and thus is limited to the process of imitation (Sultan et. al, 1990). In our case, the Bass model is better suited to our purposes since it explicitly distinguishes between the level of innovation and imitation (Guidolin et. al., 2010), so it provides the means to an interpretation of its parameters which can be used in further studies (e.g. diffusion models can be built that explain both the innovation and imitation). Besides, the Bass model has been used in sales forecasting (Lim et. al, 2003) since its parameters often show specific characteristics (similar patterns) for similar industries (e.g. parameters of innovation and imitation are similar for almost all the services in telecommunications industry). When introducing a new service, one can predict the diffusion of the service by using Bass model parameters for similar services for which we already have diffusion history data. In our case, we have Bass model estimates for broadband Internet access.

The second part of our simulation model represents UTAUT model which calculates the behavioural intentions of a person, depending on the estimated parameters of the UTAUT model, and policies which influence users' decisions. Characteristics of this model were already explained above.

Bass Model

The Bass model uses two parameters, coefficient of innovation (p) and coefficient of imitation (q). It can be used to predict the adoption on the basis of the adoption of other products, which are similar to the observed product or service.

The Bass model derives from the premise that the conditional likelihood of adoption of a randomly chosen consumer at time t, given that the adoption has not yet occurred, is a linear function of the number of previous adopters. This can be represented as a hazard function of Eq. (1)

$$f(t)/[1-F(t)] = p + (q/m)Y(t) \quad (1)$$

where m represents the total number of potential adopters, p the coefficient of innovation and q the coefficient of imitation. F(t) and f(t) are cumulative and non-cumulative proportions of adopters at time t, Y(t) is the total number of adopters by time t. The diffusion rate S(t) at time t can be derived from (1):

$$S(t) = pm + (q-p)Y(t) - (q/m)[Y(t)]^2 \quad (2)$$

Within the microsimulation model, the Bass model is implemented by a set of calculations. Firstly, the adoption rate (i.e. the number of adopters in the next time interval) is calculated according to (2):

```
AdoptionRate = (p * PopulationSize + (q - p) * Per-
sons.Adopters() - (q/PopulationSize) * Persons.Adopters() *
Persons.Adopters())/Persons.PotentialAdopters()
```

In our simulation runs, the default p and q values are the same as for broadband adoption in Italy (but they can be changed interactively in our microsimulation tool.) Once per year, each household evaluates its position and decides about the adoption according to the algorithm:

```
for(int i=1;i<=AdoptersNow; i++) {
   do {
        thisOK=false;
        Adopter=uniform_discr(0,((int)PopulationSize)-1);
        if (!Persons.get(Adopter).Adopter) {
             Persons.get(Adopter).Adopter=true;
             thisOK = true;
        }
   } while (!thisOK);
}
```

In the above algorithm, the main loop repeats as many times as there are new households to adopt DTV. Next, the algorithm randomly chooses the household which adopts DTV. Besides the Bass model itself, we added also the rule about the switchoff year:

```
if (YearsToSwitch==0) {
   AdoptersNow = Persons.PotentialAdopters();
} else {
   AdoptersNow = Math.round(AdoptionRate);
}
```

The rule says that the adoption rate should be calculated according to the Bass model, except when the switchoff date is closer than 1 year, when all households which did not adopt DTV yet should change their technology.

UTAUT Model Implementation

According to the findings of the regression models described in Section 3, the following formula is used in the simulation to calculate the use behaviour variable for each household:

```
1.023 + 0.212 * (facilitating_conditions +
get_Society().FacilitatingConditions((int)get_Society().Curre
ntYear))
+ 0.291 * (performance_expectancy +
get_Society().PerformanceExpectancy((int)get_Society().Curren
tYear))
+ 0.229 * privacy_and_security
```

According to the above, the factors PE (performance expectancy) and FC (facilitating conditions) are firstly estimated from the values of variables for each household. These values are then affected by policies Facilitating Conditions (FC) and Performance Expectancy (PE) which can have different intensities each year. The intensity is measured by one step on a Likert (1936) scale – one step means that a household would be influenced by a policy measure in such a way that it would respond by one step score on average to the questions within the questionnaire which constitutes a specific factor. An assumption is used in the model that a person becomes a service user if she scores 4 or 5 in behavioural intentions (on the Likert scale).

Scenarios

In this section a series of alternative reference scenarios are presented. They are not intended to be predictive, but only to investigate the mechanisms of our model. All scenarios refer to the Italian situation (Treré and Sapio, 2008), with initial DTV take-off at the beginning of 2004 and switch-off at the end of 2012. The impact of three different policies and combinations of them has been tested:

- *Policy PE:* communication campaign to increase the user's performance expectancy from DTV services.
- *Policy FC:* user support (through call centres and similar) to create facilitating conditions.
- *Policy PS:* communication campaign or security interventions on applications to increase the users' perceived security.

The simulation model allows the impact of two different intensity levels for each policy to be tested (level 1 = moderate, level 2 = strong). Policies based on effort expectancy (e.g. design of DTV applications with increased usability) have not been considered, since the effort predictor had a non-significant impact.

In figure 2 the DTV adopters' curve as generated by the microsimulation model is presented. The curve stays unchanged after considering the impact of the different policies since it only depends on the Bass model parameters (p=0.014, q=0.726) (Turk & Trkman, 2009).

The year 2004 was the date of the first introduction of terrestrial digital television in Italy. The adoption rate remains slow during the first years and becomes increasingly quicker when the switch-off date gets closer. An intrinsic characteristic of the model is that all users (189) become DTV adopters from 2013 when analogue broadcast will no longer be in operation.

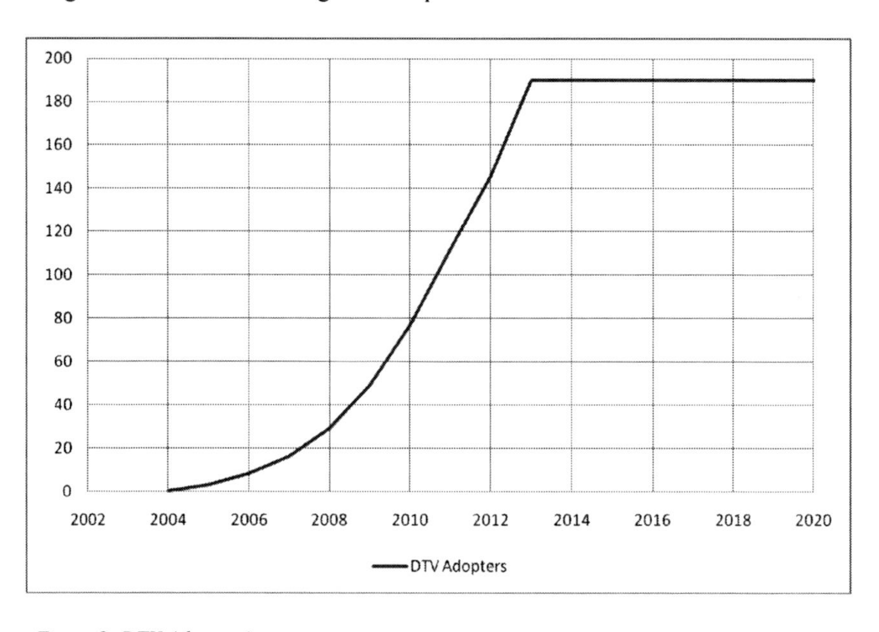

Figure 2: DTV Adopters' curve

Scenario 1: No Policies

The use behaviour (number of users of the service T-bollettino) without any governmental policies is shown in Figure 3. The use behaviour has been obtained by the item of the questionnaire B21 "intention to use the T-bollettino service". The bold line represents the mean value, whereas the thin lines refer to minus and plus three times the standard deviation. In this scenario the growth is quite regular, slower during the first years and somehow quicker from 2009, without any singularities. The regime value is slightly below three quarters of the total number of DTV users and it is reached after the switch off in 2013.

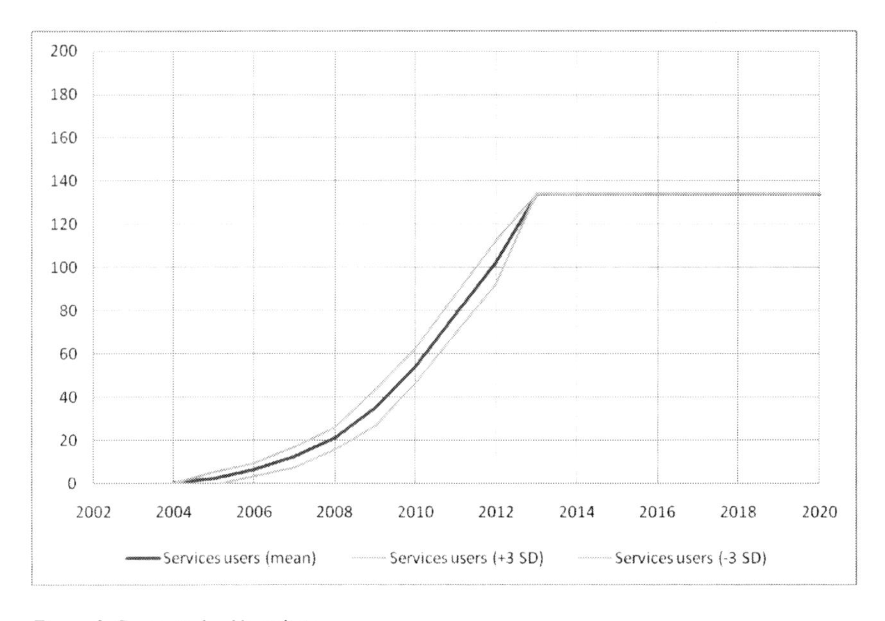

Figure 3: Scenario 1 – No policies

Scenario 2: Policy PE 2011-2013 (Level 1)

Figure 4 shows a scenario where a communication campaign to showcase the benefits of DTV payments to citizens is implemented from the year 2011 to the year 2013. A confrontation with scenario 1 (no policies) denotes an increase of the service users during the implementation of the campaign, up to a spike of nearly 85%. A longer duration of the communication campaign generates a longer lasting peak of users, yet it is hardly achievable in a real situation, due to the high costs of national campaigns.

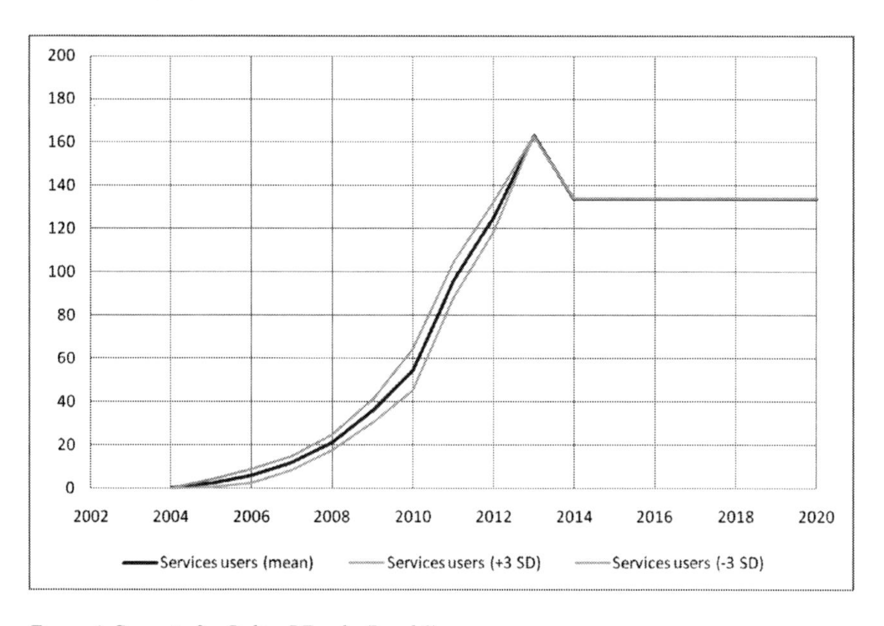

Figure 4: Scenario 2 – Policy PE only (Level 1)

Scenario 3: Policy FC 2011-2015 (Level 1)

Figure 5 shows a scenario where a policy of user support (through call centres and similar) is implemented from the year 2011 to the year 2015 at level 1 (moderate intensity). Here again there is an increase of the service users during the implementation of the policy, with comparable effects with the communication campaign policy.

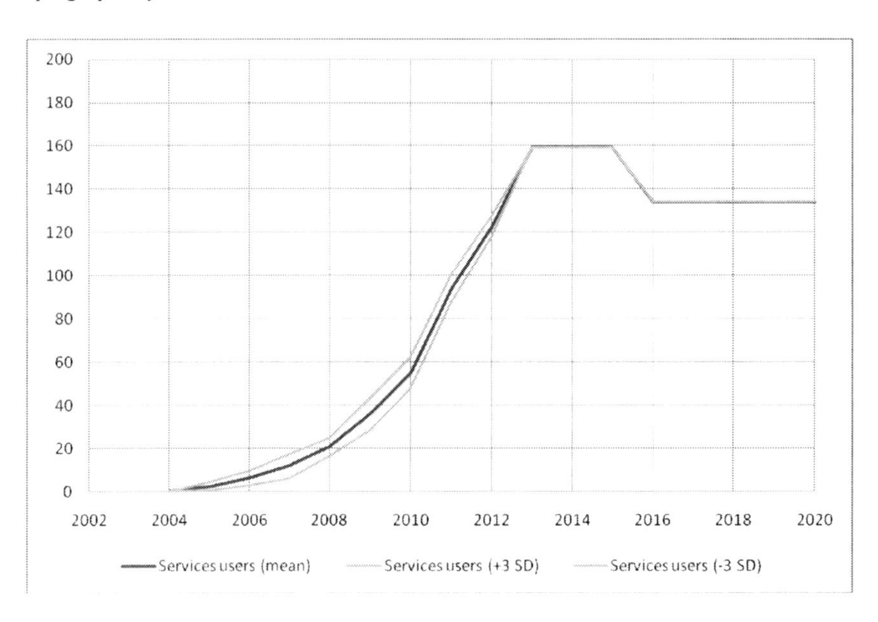

Figure 5: Scenario 3 – Policy FC only (Level 1)

Scenario 4: Policy FC 2011-2015 (Level 2)

By raising the intensity level of the FC policy to a strong user support we obtain figure 6, showing an increase of about 10% if compared to the moderate level. In a real implementation this policy could be easily sustained through a larger number of years: the presence of call centres to support consumers is common practice both in the private and public sector.

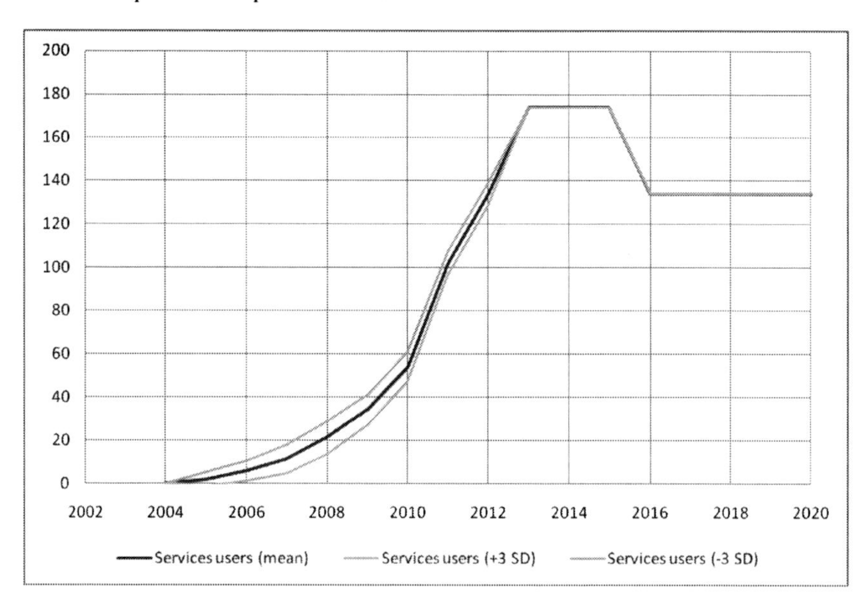

Figure 6: Scenario 4 – Policy FC only (Level 2)

Scenario 5: Policy PS 2011-2018 (Level 2)

Figure 7 highlights a scenario where policies to increase the perceived security are implemented from 2011 to 2018. These policies seem to be the most effective, pushing the number of adopters of interactive services to the highest values.

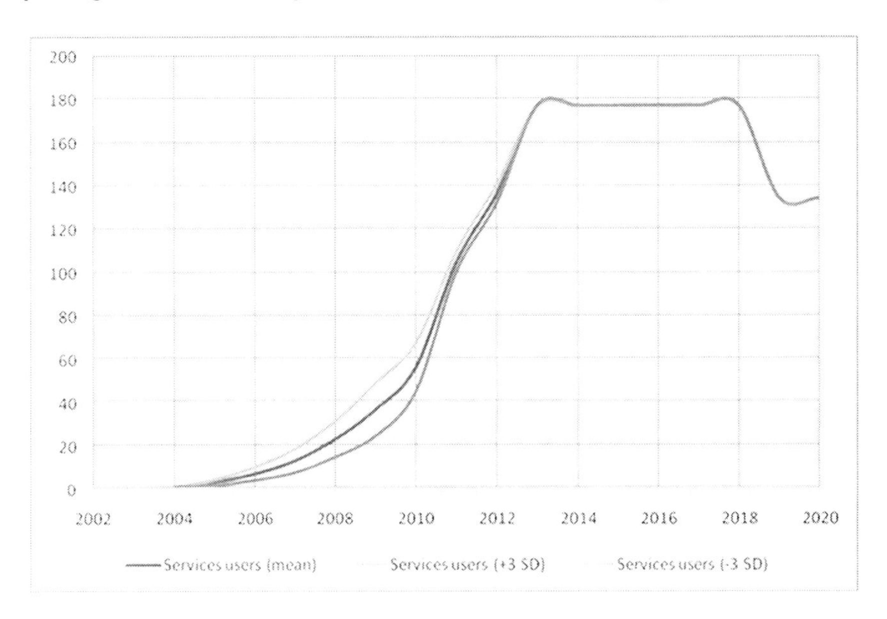

Figure 7: Scenario 5 – Policy PS only (Level 2)

Scenario 6: Combination of Policies PE 2012-2013 (Level 2) and FC 2011-2018 (Level 2)

Figure 8 shows a scenario where a combination of two different policies (PE and FC) is implemented in a realistic situation, where a communication campaign is financed by the national government for a couple of years around the switch off date with a strong level of intensity and a constant support to users through call centres and similar is provided for many years. The number of users who try the T-bollettino service goes to its upper limit in conjunction with the simultaneous action of the switch off, the campaign and the support, falling back to the regime value in two steps towards the end of the implementation of the policies.

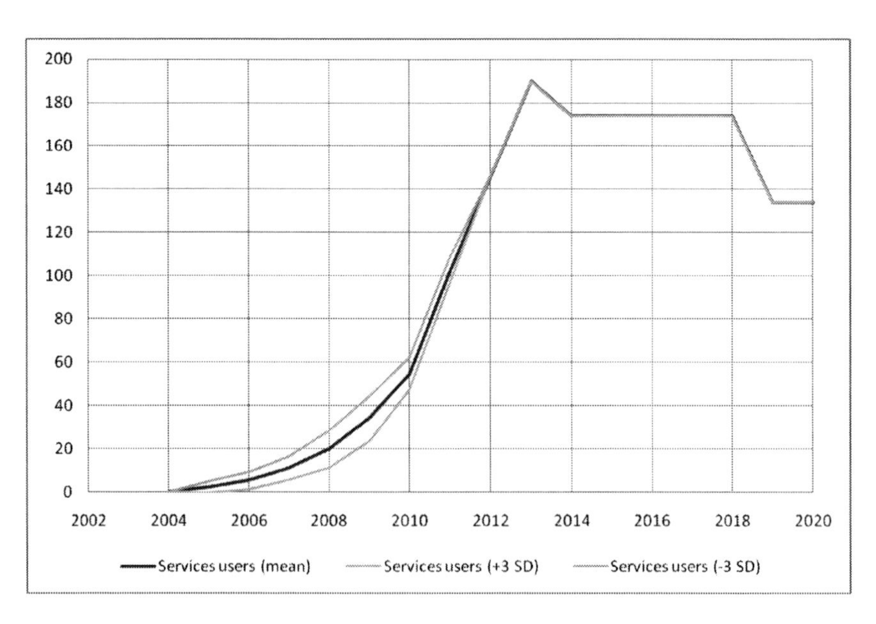

Figure 8: Scenario 6 – Combination of policies PE (Level 2) and FC (Level 2)

Scenario 7: Combination of Policies PE 2012-2013 (Level 2), FC 2011-2018 (Level 2) and PS 2011-2018 (Level 2)

By adding the PS policy we obtain scenario 7 in figure 9. The adopters of the T-bollettino service raise with the same shape during the intervening time period, but the peak reached is higher in this combination scenario.

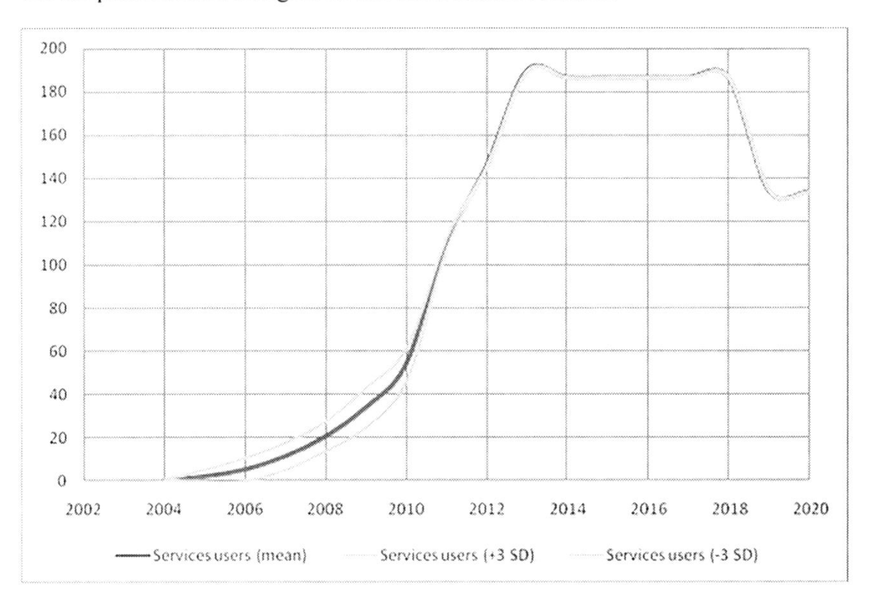

Figure 9: Scenario 7 – Combination of policies PE (Level 2), FC (Level 2) and PS (Level 2)

Conclusions

The main aim of the present study was to investigate the user experience and the usage behaviour during the diffusion of a new emerging technology (T-bollettino) that allowed payments by DTV. A multidisciplinary approach, in particular the framework provided by the human factors discipline, was adopted in order to focus the attention on the variables affecting both usage and user experience aspects. In such a perspective a predicting model of ICT user acceptance, the UTAUT model, was applied to recognize those factors directly affecting the usage. After that step, microsimulation was implemented to develop the diffusion model of payment services through DTV.

The methodology adopted in the present study was innovative approach (Sapio et al, 2010) and in the present investigation it progressed through the following main steps by:

- analysing specific user acceptance aspects for payment services (1);
- structuring the data collected in the Italian pilot study "Services for citizens via DTV" to feed the microsimulation model (2);
- identifying the most relevant factors affecting service usage adopting the UTAUT model as a theoric framework (3);
- building up the microsimulation model (4);
- generating scenarios about citizens' adoption and use of DTV services for payments (5).

Given the existing general policy constraints, different strategic scenarios on citizens' adoption and use of DTV services for payments were envisaged by depending upon governmental decisions, i.e. public communication campaigns and user support. The following main indications were extracted from the scenarios.

- Without any kind of policy, the regime value of people adopting payment services is reached after the switch off date and its value is slightly below three quarters of the total number of DTV users.
- The main effect of the realisation of a moderate intensity communication campaign to showcase the benefits of DTV payments to citizens is an increase of the service users during the implementation of the campaign, up to a spike of nearly 85%. A longer duration of the communication campaign generates a longer lasting peak of users. However this long duration of the national campaign is hardly achievable in a real situation, due to the high costs. A similar effect is produced by a moderate intensity policy of user support (through call centres and similar).
- By raising the intensity level of the user support policy to a strong user support we obtain an increase of about 10% of payment services adopters in

comparison to the moderate level. In a real implementation this policy could be easily sustained through a larger number of years: the presence of call centres to support consumers is common practice both in the private and public sector.

- A combination of two different policies where a communication campaign is financed by the national government with a strong level of intensity and a constant support to users through the provision of call centres produces a positive effect on the numbers of adopters. In fact, in conjunction with the simultaneous action of the switch off, the number of users who try the service goes to its upper limit.
- Specific policies to increase the perceived security seem to be very effective pushing the number of adopters of interactive services to the highest values. The reached value of adopters is higher combining these specific security policies with a national communication campaign and setting up a strong user support policy.

Of course these indications about the diffusion of interactive services allowing payment through DTV are related to the limited context of the Italian field study. However, final results and achievements have shown the methodology in itself is adequate for the study applied to the residential population. They also encourage future developments in the utilisation of the UTAUT model and its expansion to include the perception of security.

References

Anderson J. E., Schwager P. H. 2003. SME Adoption of Wireless LAN Technology: Applying the UTAUT Model. Information Systems Journal, 1, pp. 39-43.

Bass F. 1969. A New Product Growth for Model Consumer Durables. Management Science, 15 (5), pp. 215-227.

Cornacchia M., Papa F., Livi S., Sapio B., Nicolò E., Bruno G. 2008a. Factors Affecting the Usage of t-Government Services: an Exploratory Study, Proceedings of ICETE 2008, International Joint Conference on e-Business and Telecommunications, Porto, Portugal, 26-29 July, pp. 315-322.

Cornacchia M., Baroncini V., Livi S. 2008b. Predicting the Influence of Emerging Information and Communication Technologies on Home Life. In: J. Felipe & J. Cordeiro (eds.). Lectures Notes in Business Information Processing, 8, Web Information Systems and Technologies, Springer, pp. 184-200.

Davis F. D. 1989. Perceived Usefulness, Perceived Ease of Use, and User Acceptance of Information Technology. MIS Quarterly, 13 (3), pp. 319-340.

Dekimpe M. G., Parker P. M., Sarvary M. 2000. "Globalization": Modeling Technology Adoption Timing Across Countries. Technological Forecasting and Social Change, 63 (1), pp. 25-42.

Foon Y. S., Fah B. C. Y. 2011. Internet Banking Adoption in Kuala Lumpur: An Application of UTAUT Model. International Journal of Business and Management, 6 (4), pp. 161-167.

Guidolin M., Mortarino C. 2010. Cross-country Diffusion of Photovoltaic Systems: Modelling Choices and Forecasts for National Adoption Patterns, Technological Forecasting and Social Change, 77 (2), pp. 279-296.

Gupta B., Dasgupta S., Gupta A. 2008. Adoption of ICT in a Government Organization in a Developing Country: An empirical study. The Journal of Strategic Information Systems, 17 (2), pp. 140-154.

Kluever K. A., Zanibbi R. 2008. Video CAPTCHAs: Usability vs. Security, Rochester Institute of Technology, Rochester, NY USA, 26 September. URL: http://static.googleusercontent.com/external_content/untrusted_dlcp/research.google.com/it//pubs/archive/35117.pdf (accessed 31 January 2012).

Korres G. M. 2008. Technical Change and Economic Growth: Inside the Knowledge Based Economy. Surrey: Ashgate Publishing.

Lim B. L., Choi M., Park M. C. 2003. The Late Take-off Phenomenon in the Diffusion of Telecommunication Services: Network Effect and the Critical Mass. Information Economics and Policy, 15, pp. 537-557.

Livi S., Papa F., Nicolò E., Cornacchia M., Sapio B., Turk T. 2010. Acceptance and Use of Interactive Digital TV Services by Citizens. Communication, Politics and Culture, 43 (2), pp. 55-69.

Marchewka J.T., Liu C., Kostiwa K. 2007. An Application of the UTAUT Model for Understanding Student Perceptions Using Course Management Software, Communications of the IIMA 93, 7 (2), pp. 93-104.

Meade N., Islam T. 1995. Forecasting with Growth Curves: An Empirical Comparison. International Journal of Forecasting, 11 (2), pp. 199-215.

Monti M. 2010. A New Strategy For The Single Market At The Service Of Europe's Economy And Society, Report to the President of the European Commission, José Manuel Barroso, 9 May.

OECD. 2009. Rethinking E-Government Services: User-Centred Approaches. http://www.oecd.org/dataoecd/2/39/44590340.pdf (accessed 31 January 2012).

Papa F., Nicolo' E., Livi S., Sapio B., Cornacchia M. 2010. Factors Affecting the Usage of Payment Services through Digital Television in Italy. Proceedings of EuroITV 2010, Tampere, 9-11 June.

Sapio B., Turk T., Cornacchia M., Papa F., Nicolò E., Livi S. 2010. Building scenarios of digital television adoption: A Pilot Study. Technology Analysis & Strategic Management, 22 (1), pp. 43-63.

Sultan F., Farley J. U., Lehmann D. R. 1990. A Meta-Analysis of Applications of Diffusion Models. Journal of Marketing Research, 27 (1), pp. 70-77.

Treré E., Sapio B. 2008. DTV in Italy. In: W. Van Den Broeck & J. Pierson (eds.). Digital Television in Europe. Brussels: VUB press.
Turk T., Trkman P. 2009. Broadband Diffusion in European OECD Member Countries – Bass Model Estimates. Presentation on the COST Action 298 and IPTS Joint meeting, at European Commission - Joint Research Centre, Institute for Prospective Technological Studies, Seville, Spain (16 February), URL: http://www.cost298.org (accessed 31 January 2012).
Venkatesh V., Davis F. D. 2000. A Theoretical Extension of the Technology Acceptance Model: Four Longitudinal Field Studies. Management Science, 46 (2), pp. 186-204.
Venkatesh V., Morris M. G., Davis G. B., Davis F. D. 2003. User Acceptance of Information Technology: Toward a Unified View, MIS Quarterly, 7 (3), pp. 425-478.
Williams P. W. 2009. Assessing Mobile Learning Effectiveness and Acceptance, A dissertation submitted to The Faculty of The School of Business of the George Washington University in partial fulfilment of the requirements for the degree of Doctor of Philosophy, 31 January.
Yee K. 2002. User Interaction Design for Secure Systems, Proceedings of the Fourth International Conference on Information and Communications Security. Singapore.
Zhou T. 2008. Exploring Mobile User Acceptance Based on UTAUT and Contextual Offering, Electronic Commerce and Security, International Symposium, 3-5 August, Guangzhou City, pp. 241-245.

About the Authors

Alberto Abruzzese is Professor of Sociology of Communication at IULM University in Milan, where he also served as Dean of the Faculty of Tourism, Cultures and Territory and pro-Rector for International Relationships and Technological Innovation. His research fields are the following: mass communication, cinema, television and new media, especially with a focus on social changes related to mass uses of media. He has been for years Full Professor of Sociology of Communication at "La Sapienza" University in Rome and at the "Federico II" University in Naples. Some of his works (also translated in French, Spanish and Brazilian) are: *Forme estetiche e società di massa* (1973), Lo *splendore della TV. Origini e destino del linguaggio audiovisivo* (1995), *Lessico della Comunicazione* (2003), *L'occhio di Joker* (2006), *Sociologie della comunicazione* (with Paolo Mancini, 2007), *Educare e comunicare. Spazi e azioni dei media* (edited by, with R. Maragliano, Mondadori, 2008), *Il crepuscolo dei barbari* (Bevivino, 2011).

Juan Miguel Aguado holds a PhD in Communication Studies at the Complutense University of Madrid (Spain) and Postgraduate in Social Research by the Polish Academy of Sciences (Warsaw). Currently he is Associate Professor of Communication Theory in the School of Communication and Information Studies at the University of Murcia (Spain). His research and publications focus on the social impact of technology, mobility and the role of experiential mediation in cultural consumption processes. Together with Inmaculada J. Martínez he has recently published *Sociedad Móvil. Cultura, identidad y tecnología* (Madrid, 2008) and *Movilizad@s. Mujer y comunicaciones móviles en la Sociedad de la Información* (2009), and in co-operation with Eva Buchinger and Bernard Scott, *Technology and social complexity* (2009). He is the Head of the Mobile Media Research Project supported by the Spanish Ministry of Innovation (CSO2009-07108).

Nello Barile teaches Media studies and Sociology of cultural processes at IULM University of Milan where he is the coordinator of the Master programme in Creativity Management. He holds a PhD in Communication sciences, resources management, and formative processes at University of Rome "La Sapienza". He has written many books, articles and essays in Italy such as the recent *Sistema moda. Oggetti strategie e simboli dall'iperlusso alla società low cost* (Milano, 2011) and *Brand new world. Il consumo delle marche come forma di rappresentazione del mondo* (Milano, 2009). He also published articles and essays in France, Brazil, and USA, such as *A knot to untie. A social History of tie between fetishism, communication and power* in C. Giorcelli and P. Rabinowitz, eds. *Habits of being* (Vol. 2). Minneapolis, MN: University of Minnesota Press 2012.

Eleonora Benecchi is a PhD candidate at the Faculty of Communication Sciences at the Università della Svizzera Italiana. Her main research object is the phenomenon of fandom as seen in an economic perspective. She is also tutor of the radio project Psicoradio and active in the Japanese animation field of studies, having published the book *Anime, Cartoni con l'anima*, dedicated to the distribution and fandom of anime.

Jakob Bjur, PhD University of Gothenburg, is Research Director at TNS SIFO (part of the Kantar Media Group), and researcher and lecturer at the department of Journalism, Media and Communication, University of Gothenburg, Sweden. His dissertation *Transforming Audiences. Patterns of Individualization in Television Viewing* maps out social and cultural change in television viewing behaviour during one decade 1999-2008. After its public defense in January 2010, Bjur was appointed Researcher in Residence by first, the Swedish Radio in Stockholm and then, the Swedish Television in Stockholm. His research within academia and industry focus future development of media and media consumption, and he is specialized in audience measurement design and its consequences.

Fausto Colombo is Full Professor of Media Theories and Media and Politics at the Faculty of Political Sciences, Università Cattolica del Sacro Cuore, Milan. His scientific activity concerns sociology of media, the social history of Italian cultural industry, with particular attention at the global development of digitalization and at the processes of social shaping of digital media. He's actually coordinator of the Italian Association of Sociologists of Culture and Media (Ais-Pic). Since 2001, he has been member of many European networks of scholars and researchers. He is author, among others, of: *La cultura sottile. Media e industria culturale italiana dall' Ottocento ad oggi* – Light culture. Media and cultural industries from the 19th century and onwards (Milano: Bompiani, 1998); *I margini della cultura. Media e innovazione – Culture at its margins, media and innovation,* with L. Farinotti and F. Pasquali, (Milano: Angeli, 2001); *La digitalizzazione dei media – Media Digitalization* (Roma: Carocci, 2007); *Broadband Society and Generational Changes,* ed. with Leopoldina Fortunati (Peter Lang, 2011). Fausto Colombo is also director at OssCom, the Centre for Media and Communication Research of Università Cattolica of Milan that was founded in 1994 in order to conduct theoretical and applied research in the field of media and cultural industries in the Italian context. OssCom is also involved in European research networks, currently "Eu Kids Online III", where it is the Italian partner, the COST Action "Transforming Audiences, Transforming Societies" (ISO906) and "Developing & Investigating Methodologies for Researching Connected Learning" (DIMRCL).

Michele Cornacchia, degree in Physics, had leading work experience in multivariate statistics techniques applied to data on atomic fission and on automatic speech recognition. Since the 1986 at Fondazione Ugo Bordoni he concerned with the issues of human computer interaction (voice interfaces) and usability in the areas of emerging telecommunication system and services. With the group of Human Factors he attended a number of national and international works and projects, i.e. a socio-organisational survey committed by the Italian Communication Ministry, some COST actions and the European Project RACE 1065 ISSUE. On 1992 he received his Master's degree and Specialisation in Organisational Science and in the next years he was involved with mediated communication and telework. Most recently, he took part in the national project "Monitoring of the Italian e.m. fields" and as well as he was the group leader for the final product evaluation in ePerSpace (IST-506775), an Integrated Project under EU FP6 dedicated to the development of integrated, personalised communication services in the home area. As for the modelling, design and execution of the performance evaluation by the end-user, he is currently managing the final phase of the EU PANDORA Project (FP7-ICT-2007-1-225387) – Advanced Training Environment for Crisis Scenarios. He is the author/contributor of many scientific papers and books.

Andrea Cuman is a PhD candidate in Sociology of Cultural Processes at the University of Sacred Heart of Milan, his project investigates the experiencing of space, and its mediated interaction through the travel guidebook at the rise of so called locative media. He is also Junior Researcher at OssCom (Research Centre on Media and Communication), where he is conducting research on corporate brands and social media, and on the creative processes in the Italian cultural industry. His research interests regard the digitisation processes of media and cultural products, with particular attention to the transformations brought by social and location based media. Among the recent publications, a book chapter (co-authored with E. Locatelli) titled *Social Networks and privacy. Construction and protection of digital self.*

Manuela Farinosi is a Post-Doctoral Researcher in the Department of Human Sciences at the University of Udine (Italy). She received a PhD in Multimedia Communication from the University of Udine in 2010. Her main research interests are focused on social media, surveillance, privacy, media activism, and alternative media.

Claudio Feijóo holds a MSc and PhD in Telecommunication Engineering and a MSc in Economics. Currently he is Professor at Technical University of Madrid (UPM) where he researches on the future socio-economic impact of emerging information society technologies, in particular, from a next generation networks, mobile and/or content perspective. He serves as Deputy Director at the Research Centre for Applied ICTs (CeDInt) at UPM. He spent two years at the Institute for Prospective Technological Studies of the European Commission researching on the future prospects of mobile content and applications. He also directed the Chair in Telecommunications Regulation and Information Society Public Policies at UPM. He participated in the information society development plans and broadband deployment strategies while being adviser for the Spanish State Secretary on Telecommunications and Information Society. He is part of the Mobile Media Research Project supported by the Spanish Ministry of Innovation (CSO2009-07108), since 2009.

Leopoldina Fortunati is Director of the PhD Program in Multimedia Communication at the University of Udine where she teaches Sociology of Communication and Culture. She has conducted several researches in the field of gender studies, cultural processes and communication and information technologies. She is the Italian representative in the COST Domain Committee (ISCH, Individuals, Societies, Cultures and Health). She is associate editor of the journal *The Information Society* and serves as referee for many outstanding journals. Her works have been published in eleven languages: Bulgarian, Chinese, English, French, German, Italian, Japanese, Korean, Russian, Slovenian, and Spanish.

Julian Gebhardt is a Social Scientist with a PhD in Media- and Communication Studies. He is a Senior Researcher and Consultant in Innovation and Design and works as a Teacher, Mentor, and Coach at the HPI School of Design Thinking Potsdam. He is a trained and certified mediator and published numerous articles and co-edited books on Digital Media, Communication and Social Interaction. Julian currently lives and works in Berlin.

Leif Kramp works as Research Coordinator of the Centre for Media, Communication and Information Research (ZeMKI) at the University of Bremen. Previously he has worked as a lecturer and research associate at the Macromedia University of Applied Sciences for Media and Communications in Hamburg, as a lecturer at the Hamburg Media School and as a research fellow at the Institute for Media and Communication Policy in Berlin. He studied Journalism, Media and Communication Science, History and Economics at the University of Hamburg and was awarded a doctorate with a thesis about television as memory machine and strategies for the television heritage management. He has written and co-edited various books about media and journalism.

Inmaculada J. Martinez holds a PhD in Communication Studies at the Complutense University of Madrid (Spain) and Postgraduate in Business Management by the Know How Business School in Madrid. She has taught advertising strategies and theory at universities in Madrid, Portugal and Brazil. Currently she is Associate Professor of Advertising Theory in the School of Communication and Information Studies at the University of Murcia (Spain). Her research and publications focus on social groups' identity, and the impact of mobility and the Internet on corporate communication and branding strategies. Together with Juan Miguel Aguado she has recently published the books *Sociedad Móvil. Cultura, identidad y tecnología* (Madrid, 2008) and *Movilizad@s. Mujer y comunicaciones móviles en la Sociedad de la Información* (2009). She is part of the Mobile Media Research Project supported by the Spanish Ministry of Innovation (CSO2009-07108), since 2010.

Andrea Miconi is Assistant Professor of Media Theory and Sociology of Cultural Processes at IULM University of Milan, Italy. His research is mainly focused on theoretical, social and political implications of network society, and on mass media history as well. He is editor of *Problemi dell'informazione*, and his last book is *Reti. Origini e struttura della network society* (Roma-Bari, 2011).

Enrico Nicolò received the Doctor Laurea degree in electronic engineering, summa cum laude, from the Università degli Studi di Roma "La Sapienza", Italy, in 1983. He is a senior engineering research scientist at FUB, where he has gained experience as a scenario methodologist, project-network simulationist, telecommunication networks scholar, electromagnetic field biohazard investigator and communication ethicist. He deals with user experience of communication technologies, with a focus on young users. His research interests include the language of photographic communication. Prior to joining FUB in 1986, he was for three years with S.E.S.A. Italia, where he worked on telephone and packet-switched data networks.

Filomena Papa degree in Electronic Engineering (with honours) in 1980. Since 1981 she is with Fondazione Ugo Bordoni (FUB). Until 1986 her research activity was in the field of television systems and she was Italian Delegate of the CCIR study group 11. In 1987 she joined the human factors group, performing experimental research in videoconferencing, multimedia systems, distance learning, distance work and telemedicine. She participated as Italian expert in EU RACE Program, in the EU Cost 212 Project and in the ETSI HF1 study group. She has been involved in several European research projects: inside the ESA Olympus Program, the EU RACE Program, the EU ACTS Program, the EU FP7 and AAL. Her current research interests include: user experience, user acceptance and ICT adoption models.

Giuseppe Richeri is Full Professor at the University of Lugano where he is director of the Media and Journalism Institute and of the China Media Observatory. His main field of research is media economy and policy and social communication history. He is PhD supervisor at Communication University of China in Beijing where he gives seminars. He is author of scientific articles and books published in many countries. The more recent of them are *Economia dei media. Aspetti generali e mercato italiano* (Roma: Laterza, 2012), *The factory of ideas,* with A. Pilati (Chinese translation) (Beijing: Communication University of China Press, 2009), *and La calidad de la television*, with C. Lasagni (Buenos Ayres: La Crujia, 2006).

Marta Roel holds a PhD in Communication Studies at the Complutense University of Madrid (Spain). She has lectured on broadcasting and digital content at universities in Spain and Italy. Currently she is Associate Professor of Broadcasting Industries in the School of Communication and Information Studies at the University of Murcia (Spain). Her research and publications focus broadcasting politics and digital content. She is part of the Mobile Media Research Project supported by the Spanish Ministry of Innovation (CSO2009-07108), since 2010.

Bartolomeo Sapio Doctor Laurea Degree in Electronic Engineering summa cum laude at the University of Rome "La Sapienza", he is a researcher and project manager with Fondazione Ugo Bordoni. He has carried out methodological research in the field of scenario analysis, developing the original methods WISE, SEARCH and GIMMICKS, and applying them to multimedia, broadband networks, the Internet, mobile services and the convergence between fixed and mobile networks. He designed and implemented SIMULAB (Scenario-engineering Interactive Multimedia LABoratory), an advanced work environment to carry out research activities in the field of Scenario Engineering. He researched the diffusion of digital television and adoption patterns of interactive services. He has participated in several International projects and was Chairman of COST (COoperation in the field of Scientific and Technical research) Action 298 "Participation in the Broadband Society". He was also in the Management Committee of COST Action IS0605 "Econ@Tel" and project manager of TETRA (TErrestrial Trunked RAdio).

Sakari Taipale is a Postdoctoral Research Follow at the University of Jyväskylä, Finland. He is currently leading a postdoctoral research project on "Geographical and Social Location in the Everyday Use of ICTs" (2011-2013) funded by the Academy of Finland. Prior to this position, Dr. Taipale worked as a Senior Lecturer in Social and Public Policy and as a principal researcher in an EU-funded FP6-project on the Quality of Life in Changing Europe. His research interests relate to digital technologies and mobility studies in relation to which he has recently published peer-reviewed journal articles such as Taipale, S. (accepted) Mobility of cultures and knowledge management in contemporary Europe, The European Review; Taipale, S. (accepted), Mobilities in Finland's Information Society Strategies from 1995 to 2010, Mobilities; and Taipale, S. (2009) Recognizing Human Culture on the Internet. The Nordic Journal of Cultural Policy, 12 (1), 73-88.

Tomaž Turk is an economist and holds a Ph.D. in Information Sciences. He is Associate Professor and researcher at the Faculty of Economics, University of Ljubljana. He holds courses on Development of Information Systems, Economics of Information Technology, Economics of Telecommunications, and Business Simulations. Currently his research work includes themes like information technology adoption, economics of information technology, communication networks management and Internet society issues. He has participated in several national and international projects and published over 50 papers/book chapters, including papers in Technology Forecasting & Social Change, Telecommunications Policy, Computer Standards and Interfaces, Mathematics and Computers in Simulation, International Journal of Industrial Ergonomics and Computer Communications. He was the Vice Chair of the European Commission funded research project COST Action 298 "Participation in the Broadband Society".

Emiliano Treré is Professor at the Faculty of Political and Social Sciences of the Autonomous University of Querétaro, México. He teaches courses on new media, social movements and qualitative methodologies. His research interests focus on alternative media, social movements, digital activism and new forms of television. His work has been published in peer-reviewed journals such as *The Journal of Community Informatics*, the *ESSACHESS Journal of Communication Studies* as well as in several edited books. He is currently working on mapping digital activism in México.

Jane Vincent, PhD FRSA, joined the Digital World Research Centre University of Surrey as Research Fellow in 2002 after 21 years in the telecommunications industry. Researching the social practices of information and communication technology users Jane's studies for industry and international academic organisations on the social shaping of technology, children's and older peoples' use of mobile phones are published widely. Her work on emotions and mobile phones is published in English, German and Russian and she is co-editor of *Electronic Emotion*, the mediation of emotion via information and communication technologies (2009), with L. Fortunati, Peter Lang Oxford.

Participation in Broadband Society

Edited by Leopoldina Fortunati / Julian Gebhardt / Jane Vincent

This series publishes peer-reviewed monographs and edited volumes by internationally renowed scholars in the field of the 'social use of information and communication technologies (mass media included)', 'communication studies' and 'science and technology studies'. It provides an editorial space specifically dedicated to the collection of work that integrates new research regarding theoretical discourse, methodologies and studies from multiple disciplines such as sociology, anthropology, psychology, geography, linguistics, information science, engeneering and more.

The editors particularly welcome texts elaborating new theories, original methodological approaches and challenges to existing knowledge. Proposals aimed at scholars, professionals and operators working in the diverse field of participation in broadband society are invited from all disciplines.

Vol. 1 Leopoldina Fortunati / Jane Vincent / Julian Gebhardt / Andraz Petrovčič / Olga Vershinskaya (eds.): Interacting with Broadband Society. 2010.

Vol. 2 Julian Gebhardt / Hajo Greif / Lilia Raycheva / Claire Lobet-Maris / Amparo Lasen (eds.): Experiencing Broadband Society. 2010.

Vol. 3 Leslie Haddon (ed.): The Contemporary Internet. National and Cross-National Studies. 2011.

Vol. 4 Hajo Greif / Larissa Hjorth / Amparo Lasén / Claire Lobet-Maris (eds.): Cultures of Participation. Media Practices, Politics and Literacy. 2011.

Vol. 5 Fausto Colombo / Leopoldina Fortunati (eds.): Broadband Society and Generational Changes. 2011.

Vol. 6 Jo Pierson / Enid Mante-Meijer / Eugène Loos (eds.): New Media Technologies and User Empowerment. 2011.

Vol. 7 Alberto Abruzzese / Nello Barile / Julian Gebhardt / Jane Vincent / Leopoldina Fortunati (eds.): The New Television Ecosystem. 2012.

www.peterlang.de